THE SCRUBBER STRATEGY:

The How and Why of Flue Gas Desulfurization

THE SCRUBBER STRATEGY:

INFORM, Inc.

author
Mary Ann Baviello

with
Alexandra S. Bowie
Lillian E. Beerman

research consultant
Sophie R. Weber

editor
Richard C. Allen

Ballinger Publishing Company
Cambridge, Massachusetts

A Subsidiary of Harper & Row, Publishers, Inc.

INFORM, Inc.
381 Park Avenue South
New York, N.Y. 10016
212/689-4040

ISBN 0-918780-19-5

Library of Congress #82-81547

Cover design by Philip A. Scheuer

THE SCRUBBER STRATEGY
Table of Contents

Preface

In this report, INFORM provides facts that will help the non-technical decisionmakers in the U.S. understand a technology that can significantly reduce the polluting effects of burning coal. Those decisionmakers include legislators, regulators and utility executives, public interest groups, concerned community organizations and environmentalists who, for much of the last decade, have been involved in the debate over the merits of the broader use of our most abundant fossil fuel—coal. The use of this resource, especially in large industrial and utility plants, has created widespread and intense public controversy.

For the past four years INFORM has turned its research capabilities to defining cleaner and more economical ways of using U.S. coal supplies. In a series of studies we have explored the state of coal gasification and liquefaction alternatives; we have assessed better ways to reclaim strip-mined coal lands in the West; we have studied the potential for using a technique called fluidized-bed combustion for cleaner burning of coal in industrial and utility operations. And in this study, and in another recently issued, we have focused on finding out what cleaning coal and using flue gas desulfurization systems (called "scrubbers") can contribute to reducing the polluting effects of burning coal in utility plants.

In all its research, INFORM takes the case study approach. We document the current effects of industrial practices and explore options for improved environmental performance. We assess current benefits and costs of using various pollution controls and resource conservation programs. We explore the problems producers and users of such techniques have experienced, and their views on future markets.

All in all, both scrubbers and coal cleaning offer exciting and important possibilities for putting more coal to work in generating power in this country more economically and still meeting critical air quality standards that have been set to protect public health. The need for accurate and clear information concerning these technologies is evident: 80% of the sulfur dioxide emissions in the U.S. now come from utility power plant operations, and over 140 existing oil-fired power plants are candidates for conversion to coal use.

We earnestly hope that this documentation of the technologies of scrubber systems along with INFORM's companion study of coal cleaning, may help government and business planners and concerned citizens chart intelligent future courses and set realistic goals for meeting our energy needs in an environmentally sound manner.

Joanna D. Underwood
Executive Director
INFORM

Acknowledgements

Many people have contributed many long hours in the arduous task of completing this book. The author and research associates wish to thank these people for their time, their talents and their patience. Foremost in this list is Perrin Stryker, whose editorial excellence brought the disparate parts into a coherent report. Perrin never let us lose our perspective, asking the pertinent, probing questions that guided us during the writing phase of the study. Our Research Consultant, Sophie Weber, helped us steer our way in this research, lending much of her legal expertise to the writing of the appendix on the Clean Air Act. Randi Ferrari was responsible for getting the project started in 1979 by developing our questionnaires, contacting corporate officials and conducting many of the interviews.

We would also like to thank all those people within the FGD industry whose contribution of facts and perspectives enabled us to clarify the uses and problems of FGD technology in this report: F.D. d'Ambrosi and William G. Henke of American Air Filter Company; Walter F. Ekstrom and Dr. Raymond E. Kary of Arizona Public Service Company; Joseph Yakunich of Babcock & Wilcox; Robert M. Sherwin, Douglas R. Longwell and Gene H. Dyer of Bechtel National, Inc.; A.J. Snider of Combustion Engineering, Inc.; Robert D. O'Hara of Duquesne Light Company; Andrew Cagnetta of Ebasco Services, Inc.; David G. Olson and James Hartman of General Electric Environmental Services, Inc.; Wayne R. Johnson of Kansas City Power & Light Co.;

Robert Van Ness of Louisville Gas & Electric Company; G.A. Wagner and O.J. Quartulli of the M.W. Kellogg Company; Svend Keis Hansen, John C. Buschmann and Steven M. Kaplan of Niro Atomizer, Inc.; David R. Webster of Nevada Power Company; Joseph R. Healey and Carlton A. Johnson of Peabody Process Systems; R. Forsythe of Pennsylvania Power Company; S.J. Kowalski of Philadelphia Electric Company; William A. Glover and Richard Jordan of Public Service of New Mexico; Robert J. Gleason and Joseph T. Maurer of Research-Cottrell; Neal D. Moore of Tennessee Valley Authority; Minesh Kinkhabwala of Thyssen-CEA Environmental Systems, Inc.; Stephen J. Pfeffer and Donald Pullman of Air Correction Division of UOP, Inc.; W.H. Lord, John T. Pinkston, Richard R. Lunt and J.S. Mackenzie of United Engineers & Constructors, Inc.; and Carter Dreves and Peter Maurin of Wheelabrator-Frye.

Another round of thanks must be given to those INFORM Project Advisors who consulted with us during several key phases of this work and whose criticisms and suggestions helped improve both its readability and its accuracy: Richard Ayres, Esq. of National Resources Defense Council; Keith Frye, Air Project Consultant with the Department of Energy; Mel Horwitch of M.I.T.'s Sloan School; Dr. Gerald Hollinden, Branch Chief of TVA's Environmental and Control Assessment; Dr. Howard Hesketh of Southern Illinois University's College of Engineering; Dr. George Preston, Director of EPRI's Air Quality Control Division; Harry Perry of Resources for the future; and Jack Flynn of International Paper Co.

This very special group of people at INFORM must also be thanked: Susan Jakoplic, whose production skills and sense of humor were invaluable; Mary Maud Ferguson, an exacting copy editor whose love for the English language improved ours immensely; Risa Gerson, who carefully took care of all references; Dick Allen who assisted in editing the manuscript; Ilene Green and Lisa Rosenfield who assisted us in many areas of our research; Pat Holmes, who expertly typed hundreds of pages of copy; and Cindy Hutton, who helped us to keep our sanity by reminding us that there was a light at the end of the tunnel.

Finally, for giving us an opportunity such as this we wish to thank INFORM's Board Members and all those who work so hard to keep the organization healthy, most especially INFORM's Executive Director, Joanna Underwood, who gave us her encouragement and extensive, valuable advice all through this effort.

Chapter 1

Introduction

What Flue Gas Desulfurization is and Why it is an Issue

The scrubber strategy can be simply stated: it is the use of flue gas desulfurization (FGD) systems, called scrubbers, to remove most of the sulfur dioxide (SO_2) from the flue gas pouring out of smokestacks of coal-burning power plants. But, as a technical process, and economically and politically, the scrubber strategy is far from simple. It has forced more and more utility power plant operators to buy equipment that can cost scores of millions of dollars and hundreds of millions more to operate and maintain over a lifetime period of 30 years. FGD suppliers have lost millions in cost overruns and some charge that utility operators—their only customers to date—have slowed FGD development by their unwillingness to risk buying advanced scrubber models.

The enforced use of scrubbers has brought utilities into sharp conflict with the U.S. government over requirements of the Clean Air Act, and it has intensified the "cost-benefits" issue as to how much industry should be asked to spend to comply with air quality standards set by the Environmental Protection Agency. The scrubber strategy not only has set utility operators against environmentalists but also unexpectedly put environmentalists in the same camp with eastern and mid-western coal operators and against western coal operators. Finally, the scrubber strategy is at the center of a continuing debate over the issue of acid rain and its effect on the environment.

For all these reasons, a study of FGD technology is significant, and especially in 1982 when the Clean Air Act that spurred this technology comes up again for review by Congress. That Act, as revised in 1977, mandates that scrubbers be installed on all coal-burning boilers built since September 12, 1978. And the scrubber is today the only way that emissions of sulfur dioxide can be reduced to meet the latest standards set by EPA under the Act. But prior to the Act, primary, federal, state and local standards have been set to protect the health of the U.S. public against concentrations of sulfur dioxide.

In 1977, the latest date for which figures are available, U.S. power plants emitted 19.4 million tons of sulfur dioxide.[1] In the air this chemical is transformed into SO_3 and SO_4, and these are held to be the direct causes of air pollution that has been linked to respiratory and heart problems and, increasingly, to the formation of acid precipitation which is affecting hundreds of rivers and lakes in the U.S. and Canada, destroying fish and other aquatic life.

There is no way to calculate how much has been lost through acid rain in tourist and sports businesses, in damage to buildings, forests and people. Scientists in Sweden, Norway, U.S. and Canada have shown that the acidity of rain during the last 30 years in north temperate zones, including the U.S., is now 5 to 30 times greater than the neutral (neither acid nor alkaline) precipitation that would be expected in unpopulated areas. In fact, in 1974 a storm in Scotland dropped rain that had the acidity of vinegar.[2] Among studies examined by congressmen considering amendments to the Clean Air Act in 1977, two have underlined the threat that sulfur dioxide poses to human health and as a cause of death, hospitalization and work absenteeism rates.[3] And a report issued in October 1981 by the National Academy of Sciences concludes:

> Although claims have been made that evidence linking power plant emissions to the production of acid rain is inconclusive, we find the circumstantial evidence for their role overwhelming. [The phenomenon of acid rain] is disturbing enough to merit prompt tightening of restrictions on atmospheric emissions of fossil fuels and other large sources.[4]

While federal and local regulations under the Clean Air Act are requiring utilities and industrial companies burning coal in new boilers to install FGD systems in order to meet the established standards, many utilities have resisted the use of this technology. Some utilities, including American Electric Power Co., the nation's largest coal-burner, have argued that the impacts of SO_2 on health are not verifiable and that the link between acid rain and coal burning has not been conclusively established. Industry spokesmen have claimed that SO_2 emission standards are too strict overall and that if regulations were relaxed, utilities could use simpler FGD systems. Moreover, utilities, which function under control of public service commissions, have traditionally tried to keep rates low to consumers and have generally insisted on using only those totally proven, reliable technologies that assure that service will not be shut off because of mechanical failures.

Utilities have also complained that the cost of the technology—including costs of delays involved in meeting regulation requirements—is too high, and makes it very difficult for them to provide electricity at the lowest possible cost to consumers. In July 1979, the chairman of the Edison Electric Institute estimated that the New Source Performance Standards would increase the size of utility bills by $50 billion between 1984 and 2020, and would increase the use of oil and gas, thereby reducing the use of coal by 120 million tons per year by 1990.[5]

In the fall of 1981, the chairman of the Business Roundtable's Environmental Task Force, Ralph E. Bailey, who is also chairman of Conoco and a vice chairman of the American Mining Congress, issued a study prepared by a consulting firm, Environmental Research and Technology, that cited "excessive costs," including the "high expenses of control technology," as the first of four major problems facing companies seeking compliance with the Clean Air Act. The three other problems cited are: uncertainty in planning under the CAA, delays in decision making by regulatory agencies, and technological burdens, without the benefits, under CAA requirements for controlling "the last increments" of pollution.[6]

The argument that the costs of scrubbers to control pollution is "excessive" has been shown to be without foundation in INFORM's 10-month study of plans to convert nine New York State oil-burning utilities' plants to coal. This analysis showed that while the cost of installing scrubbers in these nine plants is large ($11.3 billion), the expense would be small compared to the savings possible in switching from oil to cheaper coal. Total net savings over the lifetime of these plants could be more than $40 billion, compared to oil burning, and they would not only produce about one third more power but sulfur dioxide emissions would be less than present emissions from oil burning. Moreover, as the study pointed out, a coal-burning utility that uses a scrubber can partially offset the scrubber's capital costs by burning cheaper coal with a higher sulfur content; in 1981 this could have cut fuel costs by 13 per cent, compared to costs of burning low-sulfur coal without scrubbers.

Ultimately, it will be up to the public and to government representatives to decide how clean the air of the U.S. should be to protect the public health and to fix standards that will accomplish this. It will be up to them to decide what price should be paid for cleaning up the air to necessary levels. But to assess the role that scrubbers may play in the process, it is essential for public debate to define as precisely as possible just how effective and reliable FGD systems are, how well they have performed to date, and at what cost.

INFORM has sought such clarification over eighteen months through case-study examinations of scrubber performance at nearly one-fourth of the 84 sites where scrubbers were used in the U.S. in 1981, and through analyses of overall studies of scrubbers done by others. INFORM's research not only clarifies the general level of performance of today's scrubbers but also analyzes some of the most effective operations. This study provides facts and specific examples that have until now not been available to the general public.

Overview of the FGD Industry

Until passage of the Clean Air Act requirements in 1970, the FGD industry had developed slowly. Even though the basic process has been studied since the turn of the century, the first scrubber systems were not built in Great Britain until the 1930's, and the technologies have developed primarily since 1960. It was not until 1967 that a full scale scrubber began operating in the United States in a coal-burning power plant owned by Union Electric Co. of Missouri. Since then, the utilities, FGD suppliers, architect/engineering firms, utilities and the federal government have all taken part in the development of this technology.

After the passage of the Clean Air Act of 1970, which gave the federal government power to set and enforce air quality standards, research on scrubbers progressed at a much faster pace. The industry started to grow rapidly as it geared up to comply with the standards established by the Environmental Protection Agency under the Act.

But for the past decade there has been a major conflict between most U.S. utilities and government over the use of scrubbers. Some utilities have tried to avoid their use, seeking other means to comply with EPA standards, such as building taller smokestacks to disperse the flue gases over a wider area and lower the recorded concentrations near the plant, burning lower sulfur coal, and using outside sources of power when weather conditions prevented adequate dispersion of the flue gases.

However, by December 1980, of the 380 coal-burning utility plants in the United States, 52 plants generating 31,109 Mw, or 13.2 percent of the U.S. coal-fired generating capacity, had installed and were operating 84 FGD systems; 139 more scrubbers at plants generating 75,737 Mw were under construction or planned.[7] By December 1990 the proportion of U.S. scrubber-equipped plants is expected to increase to 29.2 percent.[8]

In 1981, of the 707 coal-fired utility boilers producing 73 Megawatts (the power output EPA has set to determine if such boilers are significent sources of pollution), only 162 were covered under the federal 1970 New Source Performance Standards. None were covered under the Revised New Source Performance Standards of 1979.

In 1980, the FGD industry was composed of 16 FGD supplying firms and 16 architect/engineering firms serving 41 utilities that used this technology. Each of these has a specific role to play. FGD suppliers are often environmental or energy companies that custom-design, assemble and install FGD systems. Utilities sometimes hire FGD suppliers directly, but usually ask architect/engineering firms to choose the supplier. Architect/engineering firms not only prepare the engineering specifications that the FGD systems are required to meet, but also solicit bids from five or six companies supplying scrubbers. Each supplier's bid includes detailed blueprints, process flow diagrams, and a cost analysis for its FGD system. The architect/engineering firm then evaluates the bid and recommends one FGD system as the most appropriate for the utility.

The 84 FGD systems operating in 1980 have been used on boilers ranging in size from 10 to 900 Megawatts. Their sizes, types and capital and operating costs vary widely depending primarily upon the size of the utility, the sulfur content of the coal burned in the boiler and the specific SO_2 regulations in the area. Capital costs reported for scrubbers have ranged from $3 million to $149 million per unit. The scrubber manufacturing industry had sales of FGD systems that totaled about $175 million in 1979 and rose to $200 million in 1980. Sales are estimated to reach $350 million in 1985.[9]

Chapter 2

Findings

The Sample

INFORM's findings are derived from the technical literature, interviews and questionnaires, and from careful analysis of 20 scrubbing systems operating in 1980 on ten power plants owned by six utilities.

These ten power plants represent a variety of settings for FGD systems and of operating conditions. Some are in industrial and urban areas of Pennsylvania and Kansas. Others are in less populated areas of New Mexico and Nevada. Some are in areas with plentiful water supplies to meet their needs; others are in arid parts of the southwest. From six of the ten power plants in its sample, INFORM gathered sufficient detailed information about their FGD operations to discuss the types of systems used and the successes and problems in meeting the standards set by EPA. These 20 FGD systems are listed in Table I.

INFORM found significant differences between these FGD systems and their successes and problems in performing what FGD systems are designated to do: remove sulfur dioxide from the flue gas emitted by coal-burning boilers. Some of these units use high sulfur coal, others burn low sulfur coal raw or coal that has been partially cleaned of sulfur and wastes. Some of these units are using old processes, dating back to the early 1970s; others use advanced processes such as magnesium-enriched lime, to clean the sulfur from the coal gas.

The age of the boilers determines what regulations their emissions have to meet. In INFORM's sample, 12 scrubber units operate on

Table I
Utilities' Scrubber Systems Surveyed by INFORM

Scrubber	*No. of Units*	*FGD Process*	*Megawatts of Boilers*
Duquesne Light Co. (western Pa.)	2	lime (magnesium-enriched)	918
Kansas City Power & Light (western Missouri & eastern Kansas)	3	limestone & lime	990
Louisville Gas & Electric (Louisville area)	6	carbide lime & dual alkali	1,559
Nevada Power (southeastern Nevada)	3	sodium carbonate	375
Pennsylvania Power (western Pa.)	3	magnesium-enriched lime	2,475
Public Service of New Mexico (north central New Mexico)	3	Wellman-Lord	1,245

boilers constructed before 1971 (or before the Clean Air Act became effective), and so are regulated only by state or local laws. Eight units in the sample operate on boilers built between 1971 and 1978 and must meet not only state and local regulations, but also the New Source Performance Standards (NSPS) established by the Environmental Protection Agency in 1971. None of the scrubbers servicing boilers in this sample are regulated by the EPA's Revised New Source Performance Standards (RNSPS) that govern the emissions limits for all boilers whose construction started after September 17, 1978.

The significance of these differences in regulations will be discussed in detail below when specific examples will be cited to illustrate them. In addition, an appendix on the Clean Air Act explains the scope and derivation of EPA's standards.

INFORM's sample of scrubbers includes a number of systems of each of the main types currently available and in use. A comparison of INFORM's sample to the total scrubber population in the U.S. (84 systems in 1980) is offered in Table II. However, INFORM's sample does include a larger number of the lime scrubbers more commonly used in the early 1970s, and somewhat fewer of the limestone scrubbers currently most popular. In many respects, the statistics agree fairly closely with those collected nationwide by a consulting firm, PEDCo Environmental, Inc., as indicated in Table II.

Table II
Comparisons Between INFORM's Sample
and 84 FGD Systems Operating in the U.S.*

	84 FGD Systems(%)	*INFORM's Sample of 20(%)*
Lime Scrubber	31	60
Limestone Scrubber	41	5
Operating before 1978	37	65
SO_2 design removal over 80%	51	85
Boilers burned coal with less than 1% sulfur content	42	40
Boilers burned coal with over 4% sulfur content	12	10
Availability over 90% in 1980	46	30

*Data from: *EPA Utility FGD Survey:* October-December 1980, Vol. 2, EPA-600/7-81-102, Jan. 1981. Prepared by PEDCo Environmental Inc. pp. 47-656

Survey of FGD Suppliers

The FGD suppliers included in INFORM's study are ten of the largest in the industry. A summary of their present and planned production, types of units offered, and when their first units began operating, is shown in Table III.

The Findings

INFORM's findings cover these two critical areas of FGD performance:

- sulfur dioxide removal; and
- availability, reliability, operability and consistency.

And these two areas of scrubber expense:

- energy used in scrubbing; and
- capital and operating costs.

INFORM's findings also clarify factors influencing choice of an FGD system and elements that contribute to its successful operation.

Table III
Production of INFORM's Sample of 10 FGD Suppliers

	Units Currently Operating	*Units Planned or Under Construction*	*Type of System Offered*	*Operation of First System*
General Electric Environmental Services	22	10	a,b,c,e, f,g,h	1972
Combustion Engineering	16	14	a,b,c,d,e	1968
Air Correction Division of UOP	15	1	a,b,i,j	1972
Research-Cottrell	11	5	b,e,h	1973
Babcock & Wilcox	9	10	a,b,h	1972
Thyssen-CEA Environmental Systems	8	1	a,c,f,i	1974
Peabody Process Systems	6	5	b,c	1975
American Air Filter	6	2	a,b	1975
Wheelabrator-Frye (MW Kellogg)	4	4	a,b,h,k	1979
Niro Atomizer	1	6	h	1981

Key:
a - lime
b - limestone
c - lime/alkaline flyash
d - limestone alkaline flyash
e - forced oxidation/gypsum-producing
f - dual alkali
g - magnesium oxide*
h - lime/spray drying
i - sodium carbonate
j - DOWA dual alkali*
k - aqueous carbonate/spray drying*

*Not in commercial operation

Part I
SULFUR DIOXIDE REMOVAL CAPABILITIES

A. When operating, the scrubbers in INFORM's sample removed sufficient sulfur dioxide to comply with the applicable federal, state and local standards for sulfur dioxide emissions from power plants. The standards for the ten power plants surveyed by INFORM varied according to when these plants were built and, for those built before 1971, where they are located.

Twelve scrubbers in INFORM's sample are at power plants built before August 1971. These power plants are subject to local standards that can be more or less strict than the federal New Source Performance Standards (NSPS) first established under the Clean Air Act in 1971. The local standards in INFORM's sample range from emissions limits for individual units of 0.6 lbs/MBtu to 1.5/MBtu.

Eight scrubbers in INFORM's sample have to meet the 1971 federal NSPS. The NSPS set a maximum limit of 1.2 lbs SO_2/MBtu of coal that can be emitted in the flue gas of those utility boilers on which construction had begun, or which had been modified, after August 1971. For all boilers subject to the NSPS, state and local emission limits cannot be less strict than the federal limit.

None of the utilities profiled by INFORM has yet had to meet the even stricter federal standards set in 1977 under amendments to the Clean Air Act. Utility boilers on which construction began after September 12, 1978, are required to meet the limits of the Revised New Source Performance Standards (RNSPS). Sulfur dioxide emissions from these boilers will have to be reduced by at least 70 percent if the flue gas contains no more than 0.6 pounds of sulfur per million Btu (see Appendix). If higher sulfur coals are used, sulfur emissions must be reduced on a sliding scale from 70 percent up to a maximum of 90 percent. However, utilities burning very high sulfur coal with a low Btu content would have to remove more than 90 percent; for example, using a 6 percent sulfur coal with 9000 Btu/lb would require 91 percent removal of sulfur dioxide in order to meet the federal emissions standard of 1.2 lbs SO_2/MBtu.

B. The 20 operating scrubbers in INFORM's sample had actual removal rates ranging from 20 to 95 percent, using coals with a sulfur content ranging from 0.6 to 4.5 lbs of SO_2/per MBtu.

Unless otherwise noted, all actual sulfur dioxide removal rates in Table IV are based on utility calculations of *average* sulfur dioxide removal rates for the year ending December 1980. Estimated rates of sulfur dioxide removal were secured from utilities that are not closely monitoring such data.

C. Half of the scrubbers in INFORM's sample are achieving actual removal rates equivalent to or higher than their design removal rates.

The range of both "design" and actual sulfur dioxide removal rates varies widely from FGD unit to FGD unit. "Design" removal rates indicate the amount of sulfur dioxide the scrubber was engineered to remove, not what was actually removed by the scrubber. The design removal rates reported by the six utilities in INFORM's sample range from 50 to 95 percent.

Rates higher than design removal rates are actually achieved by utilities, such as Duquesne Light and Pennsylvania Power, that use chemical additives, such as magnesium oxide, or by additional effective maintenance of the system. However, one utility in INFORM's sample (Louisville Gas & Electric) has locally available a "carbide lime" that enabled four of this utility's five lime scrubbers to exceed their design removal rates of 85 percent; three scrubbers during 1980 actually removed 86, 87 and 90 percent of the sulfur dioxide and a fourth, started up in December 1980, removed from 90-95% during 1981.

D. The sulfur content of the coal can be as important as the applicable emissions standards in determining how much sulfur a scrubber will be called upon to remove.

For example, both Pennsylvania Power and Public Service Company of New Mexico are subject to strict state emissions standards: 0.6 lbs and 0.65 lbs/MBtu respectively. Pennsylvania Power, which uses high sulfur coal of over 3 percent, must achieve an 80 percent sulfur dioxide removal rate, or better, to meet local state standards. Public Service of New Mexico, on the other hand, which uses a low, 0.8 percent sulfur coal, needs to achieve only a 60 percent removal rate on two of its units to meet the state's standards (see Table IV).

In some cases, where high sulfur coal is burned and local emission(s) standards are even stricter than the strictest federal standards, an FGD system alone may not be able to meet the standard. Such strict local standards may require over 95 percent sulfur reduction or emissions of

Table IV
Average SO_2 Removal Rates of 20 FGD Units for 1980

Company Unit	*Startup Date*	*Sulfur Content of Coal(%)*	*Emission Limits*		*SO_2 Removal(%)*		
			State(lbs/MBtu)	*EPA*	*Design*	*Needed to Comply**	*Actual*
Duquesne Light							
Elrama 1-4	1975	2.3	0.6		83	85	86
Phillips 1-6	1973	2.3	0.6		83	84	87
Kansas City Power & Light							
Hawthorne 3	1972	2.5	6.1 Station limit		70	0	70†
Hawthorne 4	1972	2.5	Station limit		70	0	70†
La Cygne 1	1973	5.0	1.5		80	87	75-80†
Louisville Gas & Electric							
Cane Run 4	1976	3.75	1.2		85	82	87
Cane Run 5	1977	3.75	1.2		85	82	90
Cane Run 6	1979	4.8	1.2		95	86	95
Paddy's Run	1973	2.5	1.2		90	72	85
Mill Creek 1	1980	3.75	1.2		85	82	90-95**
Mill Creek 3	1979	3.75	1.2	NSPS	85	82	86
Nevada Power							
Reid Gardner 1	1974	.65	1.2		90	0	20-90†
Reid Gardner 2	1974	.65	1.2		90	0	20-90†
Reid Gardner 3	1976	.65	1.2	NSPS	85	0	63.38§
Pennsylvania Power							
Bruce Mansfield 1	1976	3.07	0.6	NSPS	92	89	89
Bruce Mansfield 2	1977	3.07	0.6	NSPS	92	89	90
Bruce Mansfield 3	1980	3.07	0.6	NSPS	92	89	95
Public Service Co. of New Mexico							
San Juan 1	1978	.8	0.65	NSPS	90	60	60
San Juan 2	1978	.8	60% SO_2 rmvl	NSPS	90	60	60
San Juan 3	1979	.8	1.2	NSPS	90	26	30

*Based on this calculation: $SO_2/MBtu = \frac{20 \times \text{Sulfur content of coal(\%)}}{\text{Btu value of coal(Btu/lb} \times 10^{-3})} = \frac{\text{Emissions-Emission standard}}{\text{Emissions}} = \%$ of SO_2 removal
†Estimated
**Estimate for period 12/80 (startup) through 11/81
§For October 1981 only

less than 0.6 lbs SO_2/MBtu. Two utilities in INFORM's sample, Duquesne Power & Light and Louisville Gas & Electric, had to clean some of the coal they burned, along with using scrubbers, in order to comply with strict local standards. (See INFORM's study, *Cleaning Up Coal: A Study of Coal Cleaning and the Use of Cleaned Coal.*)

E. In some cases, even where utilities can meet sulfur dioxide emission(s) standards without using scrubbers, they still need scrubbers to comply with other local standards.

Nevada Power's scrubbers in INFORM's sample are legally subject to state emission(s) standard of 1.2 lbs SO_2/MBtu. Because this utility buys coals with low sulfur content (varying from 0.3 to 1.3 percent sulfur), it can readily meet this standard using scrubbers while achieving sulfur dioxide removal rates of anywhere from 20 to 90 percent. However, Nevada Power is also subject to a state emission standard for particulates and must install and keep its scrubbers on line as the only means of meeting this standard.

Another utility studied by INFORM, Kansas City Power & Light, has had to acquire scrubbers because of difficulties in both coal sulfur content and emission standards. KCP&L's plants in Missouri operate under a lenient "station limit" that covers the combined emissions of all five of its Hawthorne boilers, instead of each one individually. Since the high station limit of 6.1 lbs SO_2/MBtu was based on the emissions of flue gas from all these boilers operating without scrubbers, KCP&L has no need of scrubbers. But it has acquired two test scrubbers to gain experience for scrubber operation that is now needed at its La Cygne plant in Kansas where a 1.5 emission standard exists and high sulfur coal (from 5 to 6 percent) is burned.

Part II
OPERATING PERFORMANCE OF SCRUBBERS

Reliability, Operability and Availability

The operating performance of scrubbers can be evaluated by three different criteria: Reliability, Operability and/or Availability.

These three indexes were devised by the firm of PEDCo Environmental Inc. for EPA to measure the operating performance of an FGD system, not its efficiency in removing sulfur dioxide from a utility boiler's flue gases. The indexes indicate how well a scrubber is performing in terms of its downtimes or failures. Failures can be caused by malfunctioning of spray pumps, valves, turbines and other mechanical defects, or by shutting down the system for general maintenance, periodic cleaning or cleaning necessitated by plugging and scaling (see Process chapter).

Each of these three indexes has its specific uses and a utility may favor one over another, depending on its experience and equipment. In INFORM's sample, the range of scrubber ratings according to these three indexes can be very wide.

Reliability

A. The reliability of scrubbers in INFORM's sample covered a wide range, from 58 percent to 100 percent. Ten, or 59 percent of the 17 scrubbers for which reliability data were available had ratings of over 90 percent.

The reliability index shows how often the scrubber is operating when it is called on to operate. It shows not only that the scrubber is in working-order (Available) but is operating when it is called upon. Utility spokesmen generally agree that Reliability is the best measure of FGD performances. Reliability tells the utility to what extent scrubber failures have caused it to shut down boilers, a fact that Operability does not show. Also, the Reliability index does not penalize a utility for a decision not to operate a boiler without the scrubber, if the boiler is only going to operate for a short period.

Operability

B. The operability of the scrubbers studied by INFORM ranged from 61 to 100 percent, with nine or 45 percent of the 20 scrubbers, having operability ratings of over 90%.

This index shows how often the scrubber and the boiler it serves are operating simultaneously. PEDCo's definition calls it "the hours the FGD system was operated divided by the hours the boiler was also operating (in a certain period) expressed as a percentage."

The Operability index may overestimate FGD performance. If a scrubber fails, the utility may choose not to operate the boiler, and thereby avoid violating the emission standards applicable to the boiler. A scrubber that fails and curtails the use of a boiler obviously is not an efficient system, and may require the utility to purchase power or to switch to other means (oil or gas) of generating the lost power, if it wishes to avoid violations of emission standards.

From the standpoint of the utility's customer, Operability index may be the best gauge of scrubber performance, for it shows how often the boiler is operating without the scrubber. In such case, the boiler is probably emitting more sulfur dioxide than the emission standards allow, and thus polluting the consumer's breathing space.

Availability

C. The availability of scrubbers in INFORM's sample ranged from 66 to 100 percent; one third, or six out of 19 units, had availability of less than 80 percent.

The availability index shows not only how often the FGD system is in working order, but whether or not it is actually used. PEDCo's definition says it is "the hours the FGD system is available for operation (whether operated or not), divided by the hours [in a certain period], expressed as a percentage." The period is usually one month.* Availability answers the question: is the system on line and able to operate?

According to some utilities, Availability may not be the best measure of scrubber performance because it fails to reflect the actual performance of the system when it is operating in conjunction with the boiler. For example, Kansas City Power & Light's Hawthorne #3 unit had only a 73 percent Availability during 1980, but scored 100 percent in both Reliability and Operability.

The Availability index, or the percentage of time a scrubber is available, also tends to overestimate FGD performance. When the boiler is not being used, there is no chance for an FGD failure, yet the system will still be "available." But a high availability is critical for a scrubber serving a baseload boiler that operates continually. If the FGD

---

* One utility (Public Service of New Mexico) in INFORM's sample defines the period as "the time the plant is generating electricity," rather than using the 30-day period established by EPA.

scrubber isn't available, the boiler may have to shut down to avoid a serious violation of emission standards.

D. Very high ratings do not always indicate a superb scrubber performance.

For example, the scrubber at Paddy's Run plant of Louisville Gas & Electric in 1980 registered 100 percent Availability and over 90 percent for both Operability and Reliability. But the boiler at that plant was used only for peak electrical demand, and was only operated in 1980 for 118 hours, or .01 percent of the year.

On the other hand, high performance ratings may be achieved by effective maintenance programs and extra scrubber equipment. For example, Duquesne Light Co.'s four Elrama units comprise a redundant system of four or five scrubber units in parallel operation. With this excess scrubber capacity, the utility can shut down and clean or repair one scrubber module, while continuing to operate the others, thereby maintaining very high percentages of all three indexes (99.9%, 99.6% and 97.8%). However, even with similar excess scrubber equipment for the six boilers at its Philips plant, defective fan systems kept all the scrubbers' performances below 80 percent.

E. The Availability figures reported by PEDCo did not always agree with the percentages given to INFORM by utilities.

Availabilities given for three units (one of Kansas City Power & Light's scrubbers and two of Louisville Gas & Electric's scrubbers) were higher, while for four units (two LG&E scrubbers, one of Pennsylvania Power's and one Public Service Co. of New Mexico scrubber) the Availability figures were lower than PEDCo's figures. However, one of these utilities (Louisville Gas & Electric) thinks that PEDCo's data is the most consistent source, even though its own calculations differ significantly from PEDCo's.

F. Using data collected by PEDCo that examined EPA scrubber performance nationwide, INFORM found that the average Availability was 83.8 percent for 61 out of 84 FGD systems operating in 1980 for which data was supplied.[1]

28 of the 61 FGD systems in PEDCo's sample, or 45 percent, had average Availabilities in 1980 of over 90 percent. Using PEDCo data on average Availability for each month, INFORM determined that from 1977 to 1980 (the years for which data from more than ten scrubbers was

available), the average availability for all scrubbers in the United States was as follows:

1977	76.4%
1978	75.9%
1979	79.8%
1980	83.8%

Note that even the low Availability figure for scrubbers in 1978 was far above the average availability for all U.S. electrical generating boilers, which in 1978 was only 63 percent.

A statistical analysis conducted for INFORM, based upon PEDCo for 84 scrubbers, showed that 10 percent of the variance in Availability of U.S. FGD systems may be explained in terms of the sulfur dioxide emission standard. This indicates that the stricter the sulfur dioxide emission standard, the higher the Availability of the scrubber. One explanation for this may be that where emission standards are higher, utilities have installed newer FGD systems with more reliable technological performance. Another explanation may be that EPA's pressure on utilities to comply with stricter sulfur dioxide emission standards, and the companies' increasing care of their equipment, are substantially improving the Availability of U.S. scrubbers.

Table V
FGD System Performance in 1980 at 6 Utilities

Unit	*Process*	*Availability(%)**	*Operability(%)*	*Reliability(%)*
Duquesne Light				
Elrama 1-4	lime	99.9	97.8	99.6
Phillips 1-6	lime	79.7	71.6	77.7
Kansas City Power & Light				
LaCygne	limestone	89.5	87.5	95.62
Hawthorne 3 (PEDCo=10-12/80)	lime	79.3	100.0	100.0
Hawthorne 4 (PEDCo=10-12/80)	lime	88.0	100.0	100.0

Louisville Gas & Electric				
Cane Run 4 (1-8/80)	lime	94.7	85.3	92.96
Cane Run 5	lime	82.7	83.7	89.05
Cane Run 6	dual alkali	85.9	82.5	88.14
Paddy's Run	lime	100.0	90.4	90.35
Mill Creek 3	lime	66.7	58.8	63.76
Mill Creek 1 (12/80 only)	lime	NA	98.5	98.5
Nevada Power				
Reid Gardner 1	sodium carbonate	93.7	92.7	92.1
Reid Gardner 2	sodium carbonate	97.5	76.9	96.38
Reid Gardner 3	sodium carbonate	87.3	86.5	86.6
Pennsylvania Power				
Bruce Mansfield 1	lime	58.2	100.0	NA
Bruce Mansfield 2	lime	76.2	100.0	NA
Bruce Mansfield 3 (9/80-12/80)	lime	68.0	100.0	NA
Public Service Co. of New Mexico				
San Juan 1	Wellman-Lord	91.6	89.2	93.0
San Juan 2	Wellman-Lord	83.5	61.1	80.9
San Juan 3	Wellman-	83.8	67.1	86.74

*Availability data supplied by utilities to INFORM. All other data is from EPA Utility FGD Survey: October-December 1980, Vol. 2, EPA-600/7-81-0125.J: Prepared by PEDCo Environmental Inc., pp. 47-656.

Consistency

A. Consistency is considered a measure of scrubber performance, but it is difficult to define and even more difficult to assess.

Consistent scrubber performance under EPA regulations requires that utilities operating coal-fired boilers on which construction began after August 17, 1971 use "continuous emission monitoring" (CEM) to measure the levels of sulfur dioxide, nitrous oxides, oxygen and gas opacity. Some state regulations that apply to such boilers also require that utilities use continuous emission monitoring (CEM). Scrubbers on boilers built before 1971 which are not required to install CEM provide regulatory agencies with analysis data on sulfur, BTU content and coal tonnage; this is used to determine sulfur dioxide emission levels.

To insure compliance with the applicable standards, all scrubbers are required to have CEM equipment to monitor emissions. The utility is expected to average the sulfur dioxide emission levels over a specific time period which varies depending on the applicable regulation. Some states require a 24-hour, others a 30-day averaging period. Boilers covered under federal NSPS of 1971 are held to a 3-hour averaging period, while boilers covered under the RNSPS of 1979 are monitored over a 30-day averaging period. (For more details on the monitoring process, see Appendix.)

There is no common standard for judging consistent scrubber performance. Emissions data gathered by the utility in this monitoring process are sent to the regional EPA offices, and also to state and local regulatory agencies, each of which determines whether or not the utility was in compliance with applicable standards. Although EPA regional offices have oversight authority, state and local agencies are responsible for enforcing the regulations. EPA will only intervene if it deems it necessary. If a regulatory agency, using its own rule-of-thumb guidelines, concludes that a utility's excess emisions were "significant," a violation is registered. Depending on the inspector's judgement, "significant" may mean a slight or large excess of emissions over the averaging period.

In order to prosecute utilities whose boilers are regulated under the NSPS, the agency must perform a "source test" in which inspectors visit the boiler site and measure emissions directly to see if the scrubber has been in violation for more than three continuous hours. The CEM data alone will not stand up in court, but CEM evidence is considered sufficient to prove violations legally for boilers regulated by RNSPS.

Emissions exceeding standards over the applicable time period may not be judged a violation. For example, a utility which is releasing excess emissions because of a scrubber's failure, may be able to show that it is operating and maintaining its scrubbers well and so may be granted a "variance" (allowed to operate the boiler without the scrubber) under which no violation will be issued despite the excess emissions. Some states and local regulatory agencies establish specific guidelines for granting a variance.

In other situations, some excess emissions may not be considered "significant" in light of a utility's past performance. For example, one Pennsylvania utility exceeded emission limits for 75 hours over a 2-year period, but because only once in this period had emissions exceeded 10 percent of the standard, they were not construed to be a significant violation.

However, a number of people at the U.S. Environmental Protection Agency suggested that bureaucratic laxity may cause violations to go unrecorded. The monitoring process continuously produces a mass of information and no one knows how much of it is interpreted or acted upon. Enforcement agencies know that litigation and the source tests required under the NSPS are both expensive and time-consuming.

B. During 1980 only four coal-fired utility plants, operating six boilers using scrubbers, were found to be in violation of sulfur dioxide emission limits by EPA.

These four utility plants were: Southern Illinois Power Coop, Marion Unit 4; Indianapolis Power and Light Company, Petersburg, Unit 3; Arizona Public Service Company, Four Corners, Units 1, 2, and 3; and Central Illinois Public Service Company, Newton, Unit 1. The only federal fine or penalty assessed against these facilities during 1980 was a $25,000 civil penalty paid by Central Illinois Public Service Company as a condition of a consent decree with EPA on December 3, 1980.

FACTORS AFFECTING THE CONSISTENCY OF SCRUBBER PERFORMANCE

A. Consistency of scrubber performance relies heavily on how well the utility operates and maintains the system.

For example, Pennsylvania Power's Bruce Mansfield plant had difficulty in getting its scrubbers to consistently meet the state emission standard during 24-hour periods. Units 1 and 2 failed to maintain adequate removal, 42 and 33 percent of the time, respectively, between March 8 and December 31, 1979. However, during the first six months of 1981, these scrubbers achieved adequate removal rates in 75 percent of the 24-hour periods. A company spokesman stated that increasing attention to the operation and maintenance of all aspects of the scrubber system, and assigning a relatively large number of employees (131 in all) to work on the three scrubber units, were primarily responsible for this improvement in scrubber consistency.

B. Utility spokesmen confirm that two measures, in addition to careful supervision and maintenance of FGD operations, can be taken by utilities to insure that they consistently meet the most stringent sulfur dioxide emission standards:

- By specifying that coal suppliers provide a fuel with a maximum allowable sulfur content which is low enough so that even when the sulfur concentration of the burning fuel peaks, the fluctuations in sulfur dioxide emissions are minimized and the chance of violations lessened.
- By specifying cleaned coal, the utility not only obtains a fuel with a considerably lower sulfur content, but also a fuel of more consistent quality (fewer peaks in sulfur content), which improves both boiler and scrubber performance (see INFORM's study: *Cleaning Up Coal: A Study of Coal Cleaning and the Use of Cleaned Coal.*)

Part III
SCRUBBER EXPENSES

Energy Used in Scrubbing

A. Electricity used by the scrubbers in INFORM's sample varies from 1 percent of boiler capacity (Louisville Gas & Electric's dual alkali system on Cane Run 6) to over 7 percent of boiler capacity used by the Wellman-Lord regenerable processes at Public Service of New Mexico's San Juan 2 boiler. (See Summary Table IX)*

*INFORM obtained data on the electricity consumed by the FGD systems at six utilities (see Table IV), not on the steam or oil energy used to reheat the flue gas or on the steam or electrical energy used to produce the sulfur or sulfuric acid by-products in the two regenerable systems (magnesium oxide and Wellman-Lord).

B. The greatest percentage—up to 85%—of the total scrubber energy requirement is the electricity consumed by the fans required to force the flue gas from the boiler through the scrubber, and by the mechanical equipment (pumps, etc.) used to circulate the absorbent through the system.

The *total* energy consumed by the different components of an FGD system is obtained from either the electrical energy produced by the utility's generators, the steam energy produced by the boilers or by oil energy used to reheat the flue gas. This total FGD energy consumption is stated in terms of the percentage of boiler capacity consumed by the system.

Total Energy Consumption of Various FGD Processes on a New 500 MW Boiler

Process	*Total Energy Consumption. (electricity, steam & oil) in millions of Btu/hour*	
	4.00% Sulfur Coal	*0.47% Sulfur Coal*
Wellman-Lord regenerable	407	244
Magnesium Oxide regenerable	381	218
Limestone wet throw-away	203	163
Lime wet throw-away	188	159
Chiyoda (gypsum producing) wet recoverable	183	128
Dual Alkali wet throw-away	140	130
Spray Drying fabric filter	31	15

Source: Bechtel National, Inc., *Economic and Design Factors of Flue Gas Desulfurization Technology* prepared for Electric Power Research Institute, CS 1323, April 1980.

C. Some variations in total energy consumption result from site-specific conditions.

For example, Pennsylvania Power's lime systems use 5 percent of the electrical energy produced by their boilers, a rate two to three times higher than the percentage of generating capacity consumed by similar systems in INFORM's sample. This is because extra energy is needed at this plant to pump the FGD wastes to a disposal site seven miles away.

Factors Affecting Energy Use in FGD Systems

A. The total energy used by an FGD system can be reduced by using untreated flue gas to reheat gas cooled by the absorber.

As the gas flows through the scrubber, it is cooled and unless reheated will condense on and corrode the lining of the stack. The amount of steam energy needed to reheat the flue gas can be reduced by diverting some of the gas flow from the boiler and using it to reheat the treated gas. Allowing part of the gas to bypass the scrubbing reduces the load on the fans and pumps and also reduces the total energy needed to operate the system. Mixing the hot untreated gas with the cooler treated gas raises the temperature of the flue gas high enough to prevent corrosion of the smokestack.

However, if this is done where it is necessary to comply with emission standards—especially if these standards require a percentage reduction of sulfur dioxide emissions—it may be necessary to increase the quantity of absorbent in the scrubber to compensate for the diverted gas flow which was not scrubbed. The cost of the extra absorbent would partially offset the energy savings of this diversion technique.

B. The age of a scrubber largely determines the amount of energy it uses.

Experience gained by operating scrubber systems has improved the efficiency of their design and operation. The newer better-designed scrubbers generally require less energy to force the flue gas through the system. In the more efficient systems, reheating the flue gas should not consume more than 1 percent of the boiler's generating capacity, while fans and pumps should consume from 1 to 1.5 percent. The fans and pumps on older, less efficient systems consume on an average of from 3 to 5 percent of generating capacity.

For example, an old system, like Louisville Gas & Electric's Paddy's Run scrubber built in 1973, used 2.8 percent of generating capacity in 1980, whereas LG&E's newer system, Mill Creek No. 1 scrubber, built in 1980, used only 1.4 percent of generating capacity.

Bechtel's Findings

Some further conclusions on the total energy used by scrubbers were developed in a 1980 study by the Bechtel Corporation, drawing on figures obtained from the Electric Power Research Institute. These included the following:

The sulfur content of the coal burned does have a direct effect on the amount of energy used by an FGD system which must meet an emission limit (such as the 1.2 lb $SO_2/10^6$ Btu standard for all boilers built after 1971). When a lower sulfur coal is used, fewer total pounds of sulfur dioxide must be removed from the flue gas, and hence less liquid absorbent has to be pumped through the system.

Marked differences in total energy consumption resulting from the use of coal with different percentages of sulfur are shown in Table VI on page 27. Note that the energy consumption per hour falls off sharply as the percentage of sulfur in the coal drops—anywhere from 7 percent to nearly half in the case of spray dryer systems.

The type of system is the most important factor determining the amount of energy consumed. Bechtel's figures in Table VI on page 27 show significant differences in the energy consumed by different types of systems. The most common scrubbers—the wet throw away systems, such as lime, limestone and the Chiyoda (gypsum-producing) system—all use about the same amount of energy. But regenerable systems, such as the Wellman-Lord and magnesium oxide units, have to consume much more energy to produce a usable by-product and to regenerate their reagents, while the wet dual-alkali process uses only one-third as much. Dry FGD systems, such as the spray drying process, which have no need for energy to reheat the flue gas, consume the least energy of all scrubbers, as little as one-sixth as much as that used by the wet-throw-away process.

Capital and Operating Costs

A. The actual capital costs of the 20 scrubbers in INFORM's sample ranged from $2.4 million (at Kansas City Power & Light) to $144 million (at Pennsylvania Power).

Adjusted capital expenditures for scrubber equipment, including modifications and sludge disposal, as reported by PEDCo for 28 utilities in 1978 (adjusted to 1980 dollars) ranged up to $149 million for a Central Illinois Public Service unit. The annual costs of a large scrubber have been reported as high as $50 million a year. Thus the total costs of a $100 million scrubber operating for 30 years at, say, $20 million a year, might well exceed $700 million.

Behind such ballpark figures lie voluminous statistics that can vary widely even for scrubbers operated by the same utility. The cost variations involve many factors: different accounting methods, different discount rates and interest charges, different boiler capacities, different conditions at operating sites, different methods of waste disposal with varying groundwater and soil conditions—as well as different construction costs (original vs. retrofitted).

The result of such diversity in scrubber economics is that direct comparisons between the costs of FGD systems are practically impossible to make without specific details on each system. INFORM sought data from public service and utility commissions, from utilities themselves, and from federal agencies, including EPA's *Utility FGD Survey* prepared by PEDCo Environmental, Inc. Utilities supplied some cost data, but the most consistent and readily available source was found in Form No. 67* originated by the Federal Power Commission and now handled by the Department of Energy. On this form utilities are supposed to report details of operating, maintenance and capital cost data to the Federal Energy Regulatory Commission; some utilities do so only partially.

A summary of the different costs of 20 scrubbers operated by the six utilities in INFORM's sample is shown in Table VI. The data, except as otherwise noted, came from FPC Form #67 showing unadjusted figures as reported by the utilities.

Actual capital costs of FGD systems can vary by several hundred percent. Note in Table VI that the lime scrubber serving Pennsylvania Power's three boilers cost $175 per kilowatt produced, or six times the $29 per kilowatt combined capital cost of the two lime scrubbers handling Kansas City Power & Light's two 85 Megawatt Hawthorne boilers. On the other hand, the capital costs of the lime scrubber serving Louisville Gas & Electric's small (72 megawatt) Paddy's Run unit were 23% greater than the cost of the lime scrubber for its 472 megawatt Mill

---

*FPC Form No. 67: Steam Electric Plant, Air and Water Quality Control Data for the Year Ending December 31, 1980, Part IV Schedule E: Stack Gas Equipment to Remove Sulfur Oxides.

Creek 3 unit. To explain such remarkable variations, a detailed examination of the cost factors included in each case would be necessary.

Table VI
Cost Data on 20 FGD Systems in 1980

Unit	Capital Costs (in thousands$)	Capital Costs ($/Kw)	Process	Annual Operating Costs (mills/Kwh)	Annual Maintenance Costs (mills/Kwh)	Adjusted† Total Annual Costs (mills/Kwh)
Duquesne Light						
Elrama 1-4	57,737	102.67	MgO/lime	4.848	.320	11.87
Phillips 1-6	51,563	126.38	MgO/lime	6.081	.570	10.27
Louisville Gas & Electric						
Cane Run 4	11,336	60.3	carbide lime	.526	.662	5.26
Cane Run 5	12,220	61.1	carbide lime	.456	.573	4.07
Cane Run 6	19,734	66.0	dual alkali	5.454	1.886	5.96
Mill Creek 1 (Dec. 1980)	19,697	55.02	carbide lime	NA	NA	NA
Mill Creek 3	18,846	42.64	carbide lime	.597	.285	3.20
Paddy's Run	3,744	52	carbide	6.51	12.15	11.28
Kansas City Power & Light						
La Cygne 1	55,760	68	limestone	.805	2.664	10.48

Hawthorne 3	2,465	29	limestone	.892	1.865	4.71
Hawthorne 4	2,465	29	limestone	.892	1.865	4.71
Nevada Power						
Reid Gardner 1	6,051	48.41	sodium carbonate	.4	.4	5.29
Reid Gardner 2	6,051	48.41	sodium carbonate	.47	.47	5.29
Reid Gardner 3	15,368	129.95	sodium carbonate	.5	.49	6.78
Pennsylvania Power						
Bruce Mansfield 1	144,375	175*	MgO/lime	2.59	2.13	10.81
Bruce Mansfield 2	144,375	175*	MgO/lime	2.5	1.11	10.81
Bruce Mansfield 3	144,375	175*	MgO/lime	2.62	.49	NA
Public Service Co. of New Mexico						
San Juan 1	47,944	132.81**	Wellman-Lord	NA	NA	15.27
San Juan 2	47,985	137.1**	Wellman-Lord	NA	NA	15.41
San Juan 3	98,806	185.03**	Wellman-Lord	NA	NA	NA

NA: Data not available from the utility, FPC or PSC
*No data available in FPC form No. 67. Data was provided by the utility or PSC
†Data from PEDCo, Inc., adjusted to reflect estimated unreported overhead, direct and fixed costs, and a 65% capacity factor
**Data from PEDCo, Inc., adjusted for unreported modification costs, sludge disposal costs, estimated unreported direct and indirect costs excluding costs for control of particulate matter, and reflecting the gross generating capacity of the unit

B. Data obtained from four utilities by INFORM show that FGD capital costs range from 22-26 percent of total plant costs. (See Table VII).

These costs are somewhat higher than those estimated by Charles Komanoff, author of "Power Plant Cost Escalation: Nuclear and Coal Capital Costs, Regulations and Economics."[2] He stated that in 1978 the capital cost of an entire FGD system generally amounted to about 21 percent of the total capital costs of a power plant, or $125/kw.

Table VII
Percent of Capital Costs Attributable to FGD on 4 Power Plants

Utility	*Type of System*	*Start of Operation*	*% of Capital Cost Attributed to FGD*
Public Service of New Mexico San Juan No. 3	Wellman-Lord	1979	26
Kansas City P&L La Cygne No. 1	Limestone	1973	22
Nevada Power Reid Gardner 3	Sodium Carbonate	1976	23
Pennsylvania Power Bruce Mansfield Nos. 1,2,3	Lime	1976-1980	24

C. Estimated capital costs of the same type of scrubber may be extremely far apart.

For example, one FGD supplier, Peabody Process Systems, claimed that at a cost of $75 per kilowatt capacity, it could supply a limestone FGD system in 1981, including every piece of equipment from gas inlet to gas outlet to dewatering equipment, but not waste disposal or detailed engineering plans or site work. Few utilities would argue that such a low cost is realistic. One FGD user, the Tennessee Valley Authority, stated that a limestone FGD system, including waste disposal (whose cost can constitute 25 percent of a system's expense) and detailed engineering plans and site work, would cost an estimated $194/Kw in 1982. Even subtracting 25 percent for the cost of waste disposal, TVA's estimate would still be $145/Kw, or nearly twice that quoted by Peabody Process Systems.

D. Operating costs in INFORM's sample range from 0.4 mills per kilowatt hour at Nevada Power to over 6 mills/kwh at Duquesne Light.

Both operating and maintenance costs of scrubbers vary widely because of site specific factors, such as atmospheric conditions, types of scrubbers, material costs, sulfur content of coal used, disposal facilities and the applicable emission regulations.

For example, Nevada Power's three sodium carbonate units using low sulfur coal under lenient local emission standards had operating costs in 1980 averaging less than 1/10 those of Duquesne Light's two magnesium/lime units using high sulfur coal and operating under strict local, state and federal standards. Operating costs of systems using regenerable systems, such as the Wellman-Lord scrubber, are likely to be much higher because of the equipment needed for regenerating the absorbent in the scrubber and the costs of producing a sulfur by-product.

Some of the cost differences are attributable to differences in costs of feed materials and chemicals. For example, Louisville Gas & Electric's scrubbers have the advantage of using a local carbide lime whose cost ($15 a ton) is about a fourth what Duquesne Light and other utilities have to pay for lime. This shows up in the operating costs of three of LG&E's carbide lime scrubbers, which are about one-tenth those of Duquesne Light's two lime scrubbers that use magnesium-enriched lime.

But many other factors can account for differences in operating costs, even those between similar scrubbers operated by the same utility. For example, the LG&E's Paddy's Run and Cane Run 4 scrubbers both use carbide lime and each requires about the same number of operating personnel. But the Cane Run 4 scrubber has feed material and chemical costs per Kwh only one-twelfth those of the Paddy's Run scrubber, and its labor costs per Kwh are only one-seventh as much. The great differences in these costs are owing to the great differences in scrubber usage: Cane Run 4 operates 70 percent of the year, while Paddy's Run scrubber operated only 1 percent of the time, being used only intermittently at periods of peak power. Therefore its costs per Kwh are inevitably much smaller.

E. Maintenance costs in INFORM's sample range from 0.4 mills per kilowatt hour at Nevada Power to 12.15 m/Kwh at Louisville Gas & Electric Co.'s Paddy Run Station.

One reason for this great spread is the unusual usage of the Paddy's Run scrubber, whose maintenance costs (material and labor) are nearly

20 times as much per Kwh as those of its Cane Run 4 scrubber. In addition the Paddy's Run scrubber hs much higher costs for labor because men and trucks are needed to haul the Paddy's Run scrubber's waste seven miles away to a disposal pond.

F. Duquesne Light Co. reported total annual costs of a scrubber in 1980 dollars of $23.9 million, including operating, maintenance and financing costs.

According to unadjusted data reported to PEDCo, annual costs of scrubbers in INFORM's sample ranged from 3.70 mills/Kwh for Louisville Gas & Electric's No. 3 unit to 16.27 mills/Kwh for Duquesne Light's Philips Units 1-6. Other total annual costs in 1980 dollars reported to PEDCo Inc. ran as low as $51,784 for each of Arizona Electric Power's two units, but when these were adjusted by PEDCo to reflect estimated related costs, the total figure came to $3.7 million a year. The highest adjusted total annual cost in 1980 reported to PEDCo was $50,794,500 for each of two Pennsylvania Power Bruce Mansfield units.

What Bechtel Found

Differences in capital costs for limestone scrubbers have been estimated by Bechtel National Inc.[3] which found that the cost of a limestone scrubber in 1978 was $157/Kw when 4.0 percent sulfur was used and $118/Kw when very low sulfur (0.48 percent) was used.

Bechtel has also estimated the capital cost differences between spray drying scrubbers and systems using regenerable chemicals, such as the Wellman and magnesium oxide systems. Using high sulfur coal, Bechtel estimates the spray drying process can save from 18.8 percent to 32.8 percent of capital costs, while for scrubbers using low sulfur coal the savings can range from 57.7 percent to 64.9 percent. Wellman Lord systems using high sulfur coal, Bechtel estimates, will cost from 18.8 percent to 37.7 percent higher than lime systems, while low sulfur coal, the capital costs range from 1.6 percent to 63 percent more than lime systems.

The factors recommended by Bechtel to be used in preparing estimated capital costs of scrubbers have been listed as follows:

1. The physical scope of the FGD system including particulate removal, energy consumption, and waste disposal;

2. the date of the estimate, including inflation allowances, interest rates, and the cost of materials and labor;
3. the design of the system;
4. thc type of coal and its sulfur content that will be used in the boiler;
5. the sulfur dioxide removal requirements;
6. the factors imposed by the utility such as a maintenance contract, equipment redundance, materials of construction and performance guarantees;
7. the location of the coal plant; and
8. the accounting procedures.

Bechtel has established some criteria for the variables in FGD costs that must be analyzed in computing both capital and annual costs of the following processes: limestone, lime, Chiyoda (gypsum-producing), dual alkali, magnesium oxide, Wellman Lord and spray drying for boilers using both low (0.48 percent) and high (4.00 percent) sulfur coal. Bechtel's cost estimates shown in Table VIII include waste disposal or regeneration facilities where applicable, but do not include particulate removal.

Part IV
SUCCESSFUL FGD SYSTEMS

Factors Influencing the Choice of an FGD System

A. A primary factor that determines which FGD system is chosen is, of course, its cost.

Utilities will generally look at the lowest projected "total cost" for a system—considering both estimated capital and annual operation and maintenance costs for the life of the system. However, a more costly, novel FGD system may be chosen to meet local emission requirements or other conditions, such as scarcity of water, lack of a waste disposal site, or the availability of a market for a sulfur or gypsum by-product.

In any case, while some FGD costs can be passed along to consumers through higher electric rates, an additional inducement to choosing the least expensive system is that it is easier to justify to public service commissions, whose concern is with the effect of utility expenditures on customer rates.

Table VIII

FGD Capital, Operating and Maintenance Costs & Revenue Requirements

Process	*Capital Cost in Dollars per Kilowatt (1978)*		*First Year O&M Costs in Million Dollars per Kilowatt-yr (1978)*		*Levelized Revenue Requirements in mills per kilowatt hour (1978-2007)*	
	4.00% Sulfur	*0.48% Sulfur*	*4.00% Sulfur*	*0.48% Sulfur*	*4.00% Sulfur*	*0.48% Sulfur*
Limestone	157	118	26.3	14.5	12.7	8.0
Lime	149	111	28.3	14.5	13.2	7.6
Chiyoda (Gypsum producing)	164	125	22.2	11.8	11.7	7.2
Dual Alkali	166	134	27.3	14.0	13.5	8.2
Magnesium Oxide	176	129	24.9	14.9	12.7	8.6
Wellman-Lord	180	127	31.3	17.3	15.0	9.1
Spray Drying	121*	47	21.8*	7.4	10.3*	3.7

Basis: Two new 500 Mw boilers; midwest location, 30 year plant life, capacity factor 70%; 85% SO_2 emission reduction with a maximum allowable emission of 1.2 lb $SO_2/10^6$ Btu and a minimum controlled emission of 0.2 lb $SO_2/10^6$ Btu

*Spray drying systems above 1% sulfur are based upon theoretical calculations from bench scale testing

Source: Bechtel National, Inc., *Economic and Design Factors of Flue Gas Desulfurization Technology*, prepared for Electric Power Research Institute, C S - 1428, April 1980

B. Most utilities still choose the traditional and most highly developed FGD systems, i.e., those that use a mixture of lime or limestone and water to remove the sulfur dioxide from the flue gas while producing large quantities of sludge as waste (see Process chapter).

Many other types of scrubbers are available today, such as those that produce a dry waste instead of a wet sludge, and those that produce a saleable product, e.g., gypsum or sulfuric acid, instead of one that is thrown away. While in the West, where water is scarce, dry-waste scrubbers are being widely used, the spray dryer scrubber has not yet been tried on high sulfur coal and the Wellman Lord and gypsum-producing scrubbers have been considered too costly to be competitive, or too untried. As one utility spokesman said, "People don't buy what they don't understand." By 1981, four out of five new FGD installations were lower-cost, enconomical limestone scrubbers.

C. The choice of a particular scrubber by an architect-engineering firm and a utility does not depend on its sulfur dioxide removal capacity alone, but can be the result of nontechnical or subjective factors.

Almost every supplier offers similar lime and limestone systems that they say can remove at least 90 percent of the sulfur dioxide and meet federal emissions standards. Consequently, the architect/engineering firm may base its choice on a preference for a particular FGD supplier, or the utility may be influenced by the supplier's experience in the field, including both the number of years it has been in the business and the number of systems it has sold; or the utility may be favorably impressed by the guarantees offered by the supplier on its system.

Guarantees vary from contract to contract. Most contracts cover the rate of sulfur dioxide removal, the use of sulfur dioxide absorbing chemical and the amount of energy consumed by the scrubber. Some suppliers also guarantee that their systems will be available to operate at least 90 percent of the time. All of these guarantees are contingent upon the proper operation of the FGD system and boiler, and the use of a coal with no more than a specified sulfur content.

Contract guarantees do not include the cost of the electricity that has to be purchased when an FGD system fails. Utilities want the most comprehensive guarantees they can obtain, while the suppliers do not want to assume too much risk by guaranteeing too much for too long.

Private insurance companies will not insure scrubbers for

guaranteed performance. However, scrubbers, along with the rest of the plant, are insured to cover damage from fire and other disasters, as well as worker injuries.

D. As of 1980, only 11 utilities in the United States had adopted novel FGD systems for their plants.

These systems include the dual alkali, sodium carbonate, Wellman Lord, magnesium oxide and spray drying systems. (See Process chapter.) The same factors that affect the utility's choice of a particular lime or limestone scrubber may also apply here, but usually these systems are chosen because of special site-specific criteria.

For example, sodium carbonate scrubbers are useful only in the arid West where evaporation exceeds rainfall. Or land may not be available at or near a power plant for waste disposal, and an FGD system may be chosen that would produce a saleable byproduct instead of waste. Of if the sulfur content of the coal burned is less than 1 percent, and the required sulfur dioxide removal rate is less than 80 percent, then dry scrubbers can be used that consume less energy and cost less than the conventional wet lime or limestone scrubbers.

While most utilities still do not accept the risk of installing novel systems in their coal plants, there are notable exceptions, such as Philadelphia Electric Co., Louisville Gas & Electric and Public Service of New Mexico. (See profiles.) But according to FGD suppliers, the uncertainty over whether a full-scale unit will operate as well as its pilot-plant prototype makes utilities skeptical about using technology that has not yet been commercially proven. Suppliers say this has slowed the development and commercialization of the more innovative FGD systems, and an executive of the Air Correction Division of UOP, Inc. put it this way:

> The utility industry is not willing to take acceptable risks in adopting new technologies and new design approaches, and therefore, the entire FGD industry has not progressed to the point where it could be if more progressive customers were purchasing units.

For a utility, a possible benefit in trying novel FGD systems has to be weighed against their higher costs and the resistance of public service commissions to granting rate increases to pay for technologies not yet commercially proven. Also, utilities must consider the possibility of equipment failure which would not be countenanced by a public service commission. Most utilities have chosen the well-established systems which they assume are more reliable.

E. The choice of low-cost scrubbers is encouraged by competition among FGD suppliers, some of whom concede that they practice underbidding to get a contract and thus artificially reduce the "total cost" of scrubbers.

One supplier stated: "The key to securing orders lies in the supplier's willingness to accept lower margins and higher risks." But by making a low bid, a supplier often does not leave enough margin to cover those costs that are difficult to project accurately, such as costs due to inflation and to purchasing FGD components from subcontractors. Another supplier noted that some suppliers "by virtue of a low price receive contracts that prove ultimately to be financially disastrous."

Many FGD suppliers have lost large amounts of money fulfilling their FGD contracts. It is not clear to what extent losses can be attributed to the scrubber contracts alone, since in published financial data, the income and expenditures relating to these contracts are not isolated from the costs of other air quality control technologies. However, Peabody International Corporation, for example, stated in its 1980 Annual Report:

> ... we suffered severe cost overruns and profit deterioration on a large sulfur dioxide (SO_2) scrubber project for a utility power plant in Colorado. During 1980, we charged $5.9 million to pretax income for these overruns... While this project has been a severe profit detractor, its technical achievement is rated high in the power generation community.

Also, Envirotech Corporation reported in its 1979 Annual Report:

> ... The substantial profit decline (of Air Quality Control Group) was primarily due to construction cost overruns on several large projects and increased expenditures associated with bidding on the emission control systems which are required for the new, large coal-fired power plants.

Primarily because of a $20 million loss that was incurred on two FGD contracts with the Tennessee Valley Authority, the Envirotech Corporation sold its two divisions that supply FGD systems to General Electric in April of 1981[4] (see General Electric Environmental Services profile).

Elements Contributing to Good Scrubber Performance

A. According to the utilities and suppliers interviewed by INFORM, the effectiveness of scrubbers in meeting sulfur dioxide emission(s) requirements is dependent on four factors:

- **the use of high quality components;**
- **operators who are highly trained;**
- **an effective maintenance program; and**
- **the use of a coal with a specified sulfur content.**

Eighty percent of the suppliers interviewed by INFORM said that the peformance of their systems could be improved by having better trained operators. Other ways cited by the suppliers to improve scrubber operations were: better maintenance programs, duplicate or standby components within the FGD systems, and the presence of a chemical engineer or chemist at the power plant. Utilities that have successful FGD systems, such as Pennsylvania Power and Duquesne Light (see profiles) have already incorporated a number of these suggestions through a sizable investment of both time and money.

However, according to two suppliers, some utilities use the FGD system as the training ground for their inexperienced workers, who, once trained, are transferred to the operation of the boiler. As one manager put this policy, "Let the kids operate the FGD system and the men operate the boilers."

B. Whether or not the system will be a success is determined by the practices, policies and expertise of the three parties concerned: the utilities, the architect/engineering firms, and the suppliers.

A utility installing its first FGD system has to rely entirely upon an architect/engineeing firm to choose an appropriate scrubber and adapt it to the power plant. However, an architect/engineering firm with little or no experience with FGD systems may be chosen by a utility simply because it is the same firm that designed, engineered and built the power plant. If the FGD specifications are prepared by such an architect/engineering firm, the scrubber may not perform satisfactorily.

C. An architect/engineering firm may prepare FGD specifications that are either too narrow or too broad.

Both utilities and FGD suppliers interviewed by INFORM pointed out that where such specifications have been too broad, systems have been built with low quality components that tend to break down and wear out after little use. Where such specifications have been too narrow, the capital cost of the scrubbers usually increases and the FGD suppliers may lose money in fulfilling their contracts. In addition, some suppliers may lose FGD contracts because they cannot adapt their system to meet these narrow specifications.

Many suppliers resent the architect/engineering firms as the "third party" which adds to an already confusing situation. Most suppliers would rather deal directly with a utility which is knowledgeable about FGD, such as Louisville Gas & Electric Company (see profile), rather than an architect/engineering firm. Since only a handful of utilities have such expertise, this is not often done.

D. An FGD system may fail to work consistently and reliability because the supplier's low bid includes low-grade equipment that breaks down or fails to achieve the performance to meet the emission standard. According to some utilities interviewed by INFORM, some suppliers sell poor equipment because they don't know enough about the chemical processes of FGD systems.

E. With the installation of each successive scrubber, a utility's FGD expertise increases and its reliance upon its architect/engineering firm decreases.

At this point, a utility is capable of suggesting design changes to its architect/engineering firm in order to improve the operation of future scrubbers. Examples of three utilities that have assumed many of the tasks normally completed by architect/engineering firms are Louisville Gas and Electric, Nevada Power, and the Public Service of New Mexico. All have FGD systems that were operating quite well in 1981.

Data Summary for FGD Systems in INFORM's Sample

Utility	Initial Operation	Process	FGD Capacity (MW)	Make-up Water (gallons)	Electric Energy Consumed (%)	SO_2 Removal Rate(%)	Sulfur in Coal(%)	Availability (%)(1980)	Capital Costs $/KW	Annual Costs mills/Kwh
Duquesne Light										
Elrama 1-4	1975	lime	510	700	2.35	86	2-2.3	99.9	102.67	5.168
Phillips 1-6	1973	lime	408	600	2.35	87	2-2.3	79.7	126.38	6.651
Kansas City Power & Light										
LaCygne 1	1973	limestone	820	1148	2.70	70÷	5-6	89.5	68	3.469
Hawthorne 3	1972	lime	85	NA	2.20	70÷	2.5	79.3	29	2.757
Hawthorne 4	1972	lime	85	NA	2.20	75-80÷	2.5	88.0		
Louisville Gas & Electric										
Cane Run 4	1976	lime	188	80	1.60	87	3.75	94.7	60.3	1.188
Cane Run 5	1977	lime	200	100	1.50	90	3.75	82.7	61.1	1.029
Cane Run 6	1979	dual alkali	299	150	1.00	95	4.80	85.9	66.0	7.34
Mill Creek 1	1980	lime	358	150	1.40	90-95	3.75	NA	55.02	NA
Mill Creek 3	1978	lime	442	300	1.60	86	3.75	66.7	42.64	.882
Paddy's Run 6	1973	lime	72	20	2.80	90	2.50	100.0	52÷	18.66
Nevada Power										
Reid Gardner 1	1974	sodium carbonate	125	155	5.00	20-90÷	0.65	93.7	48.41	.8
Reid Gardner 2	1974	sodium carbonate	125	155	5.00	20-90÷	0.65	97.5	48.41	.94
Reid Gardner 3	1976	sodium carbonate	125	155	5.00	68(10/81)	0.65	87.3	126.95	.99
Pennsylvania Power										
Bruce Mansfield 1	1976	lime	825	NA	5.00	89	3.07	58.2	175*	4.72
Bruce Mansfield 2	1977	lime	825	NA	5.00	90	3.07	76.2	175*	3.61
Bruce Mansfield 3	1980	lime	825	NA	5.00	95	3.07	68.0	175*	3.14
Public Service of New Mexico										
San Juan 1	1978	Wellman-Lord	361	365	6.92	60	0.8	91.6	132.81	NA
San Juan 2	1978	Wellman-Lord	350	365	7.14	60	0.8	83.5	137.1	NA
San Juan 3	1979	Wellman-Lord	534	NA	4.68	30	0.8	83.8	185.03	NA

*No data available in FPC form No. 67. Data was provided by the utilities or PSC
÷Estimated

TABLE X SUMMARY DATA ON 81 OPERATING FGD SYSTEMS

Company	Unit	Startup	Capacity (Mw)	Type of System	Sulfur Content of Coal (%)	Boiler Energy Con-sumer (%)	Makeup Water Con-sumed (gpm)	Design Removal Rates (%)		Average (%)		
								SO_2	Partic-ulates	Avail-ability	Reli-ability	Oper-ability
Alabama Electric Cooperative	Tombigbee 2	9/78	179	Limestone	1.15	3.1	140	59.5	99.3	84.6	49.1	59.8
Alabama Electric Cooperative	Tombigbee 3	6.79	179	Limestone	1.15	3.1	140	59.5	99.3	97.1	87	64.4
Arizona Electric Power Cooperative	Apache 2	1/79	98	Limestone	0.50	4.1	1840	42.5	99.5	80.8	79.7	52.2
Arizona Electric Power Cooperative	Apache 3	4/79	98	Limestone	0.50	4.1	1840	42.5	99.5	75.4	94.8	71
Arizona Public Service	Cholla 1	12/73	119	Limestone	0.50	3/4	120	58.5	80	94	92.6	92.4
Arizona Public Service	Cholla 2	6/78	264	Limestone	0.50	---	120	75	99.7	---	---	---
Arizona Public Service	4 Corners 1	11/79	175	Alkaline Fly Ash	0.75	---	---	80	99	---	---	---
Arizona Public Service	4 Corners 2	11/79	175	Alkaline Fly Ash	0.75	---	---	80	99	---	---	---
Arizone Public Service	4 Corners 3	11/79	229	Alkaline Fly Ash	0.75	---	---	80	99	---	---	---

Company	Unit	Startup	Capacity (Mw)	Type of System	Sulfur Content of Coal (%)	Boiler Energy Consumed (%)	Makeup Water Consumed (gpm)	Design Removal Rates (%)		Average (%)		
								SO_2	Particulates	Availability	Reliability	Operability
Basin Electric Power	Laramie River 1	7.80	570	Limestone	0.81	---	269	90	99.6	100	100	99.4
Big Rivers Electric	Green 1	12/79	242	Lime	3.75	---	---	90	99	69.4	79	66.1
Big Rivers Electric	Green 2	12/80	242	Lime	3.75	---	---	90	99	---	---	---
Central Illinois Light	Duck Creek 1	8/78	416	Limestone	3.66	2.9	600	85	ESP	52.1	55.7	51.6
Central Illinois Light	Newton 1	12/79	617	Dual Alkali	2.25	2.4	---	90	99.5	52.2	44.2	38
Columbus & Southern Ohio Electric	Conesville 5	2/77	411	Lime	4.67	3.9	500	89.5	---	56.2	53.5	50.4
Columbus & Southern Ohio Electric	Conesville 6	7/78	411	Lime	4.67	3.9	500	89.5	---	71.5	54.5	52.4
Colorado Utility & Electric Association	Craig 1	8/80	447	Limestone	0.45	5.4	316	85	99.8	62.7	78.7	56.8
Colorado Utility & Electric Association	Craig 2	5/80	400	Limestone	0.45	1.5	316	85	99.8	69.2	67.9	66.1

Company	Unit	Startup	Capacity (Mw)	Type of System	Sulfur Content of Coal (%)	Boiler Energy Consumed (%)	Makeup water Consumed (gpm)	Design Removal Rates (%)		Average (%)		
								SO_2	Particulates	Availability	Reliability	Operability
Commonwealth Edison	Powerton 51	6/81	450	Limestone	3.53	5.6	560	75.5	ESP	---	---	0
Cooperative Power Association	Coal Creek 1	8/79	327	Alkali Fly Ash	0.63	---	---	54	99.5	94.2	100	61
Cooperative Power Association	Coal Creek 2	9/80	327	Alkali Fly Ash	0.63	---	---	54	99/5	---	---	61.2
Delmarva Power & Light	Delaware City 1	5.80	60	Wellman-Lord	7.00	---	---	90	90	83.5	83.1	86.5
Delmarva Power & Light	Delaware City 2	5/80	60	Wellman-Lord	7.00	---	---	90	90	82.5	82.5	82.5
Delmarva Power & Light	Delaware City 3	5/80	60	Wellman-Lord	7.00	---	---	90	90	66.0	65.6	65.6
Duquesne Light	Elrama 1-4	10/75	510	Lime	2.20	3.5	600	83	99	99.9	99.6	91.5
Duquesne Light	Phillips 1-6	6/74	408	Lime	1.92	3.4	700	83	99	73.1	74.9	67.8
Southern Illinois Power Cooperative	Marion 4	6/79	173	Limestone	3.75	2.3	---	89.4	ESP	84.5	72.6	82.3
Indianapolis Power & Light	Petersburg 3	12/77	532	Limestone	3.25	2.4	882	85	99.3	---	---	---
Kansas City Power & Light	Hawthorn 3	11/72	90	Lime	3.00	2.2	---	70	99	79.3	100	84.1

Company	Unit	Startup	Capacity (Mw)	Type of System	Sulfur Content of Coal (%)	Boiler Energy Consumed (%)	Makeup Water Consumed (gpm)	Design Removal Rates (%)		Average (%)		
								SO_2	Particulates	Availability	Reliability	Operability
Kansas City Power & Light	Hawthorn 4	8/72	90	Lime	3.00	2.2	---	70	99	88.0	100	74.4
Kansas City Power & Light	LaCygne 1	6/73	820	Limestone	5.39	2.9	1148	80	99.5	89.7	95.0	76.0
Kansas Power & Light	Jeffrey 1	8/78	540	Limestone	0.32	---	557	80	99	100	100	100
Kansas Power & Light	Jeffrey 2	1/80	490	Limestone	0.30	---	---	80	ESP	100	100	100
Kansas Power & Light	Lawrence 4	1.77	125	Limestone	0.55	---	---	73	98.9	98.6	99.6	99.0
Kansas Power & Light	Lawrence 5	4/.78	420	Limestone	0.55	---	---	52	98.9	100	100	100
Kentucky Utilities	Green River 1-3	6/76	64	Lime	4.00	3.1	75	80	99.5	76.1	86.6	80.8
Louisville Gas & Electric	Cane Run 4	8/77	188	Lime	3.75	1.5	100	85	99	93.0	88.1	67.4
Louisville Gas & Electric	Cane Run 5	7/78	200	Lime	3.75	1.5	100	85	99	75.5	71.4	81.7
Louisville Gas & Electric	Cane Run 6	4/79	299	Dual Alkali	4.80	1.0	278	95	99	93.0	88.1	67.4
Louisville Gas & Electric	Mill Creek 1	4.81	358	Lime	3.75	1.4	150	85	---	45.1	98.5	98.5
Louisville Gas & Electric	Mill Creek 3	3/79	442	Lime	3.75	1.6	150	85	99	7.37	63.8	64.8
Louisville Gas & Electric	Paddy's Run 6	4.73	72	Lime	2.50	2.8	50	90	99.1	100	90.4	63.6

Company	Unit	Startup	Capacity (Mw)	Type of System	Sulfur Content of Coal (%)	Boiler Energy Consumed (gpm)	Design Removal Rates (%)			Average (%)		
							SO_2	Particulates		Availability	Reliability	Operability
Minnkota Power Cooperative	Milton R. Young 2	6/78	185	Alkaline Fly Ash	0.70	1.6	700	36.13	99.6	52.4	51.7	47.6
Minnesota Power & Light	Clay Boswell 4	4/80	475	Alkaline Fly Ash	0.94	1.3	895	84.55	99.73	100	100	76.6
Montana Power	Colstrip 1	11/75	360	Alkaline Fly Ash	0.77	3.3	370	60	99.5	82.7	---	88.4
Montana Power	Colstrip 2	10/76	360	Alkaline Fly Ash	0.77	3.3	370	60	99.5	93.8	---	---
Monongahela Power	Pleasants 1	3.79	618	Lime	3.70	---	---	90	99.55	0	3.5	24.4
Monongahela Power	Pleasants 2	10/79	618	Lime	4.50	---	---	90	99.55	---	3.5	24.4
Nevada Power	Reid Gardner 1	4/74	125	Aqueous Sodium Carbonate	0.50	5	155	90	97	84.2	81.1	78.6
Nevada Power	Reid Gardner 2	4/74	125	Aqueous Sodium Carbonate	0.50	5	155	90	97	88.4	86.2	82.6
Nevada Power	Reid Gardner 3	7/76	125	Aqueous Sodium Carbonate	0.50	5	155	85	99	84.1	84.7	81.9
Northern Indiana Public Service	Dean H. Mitchell II	6/77	115	Wellman-Lord	3.50	12.1	---	0-	98.5	62.4	56.7	55.6
Northern States Power	Riverside 6-7	12/80	110	Spray Drying	1.20	---	---	90	99.5	---	---	---

Company	Unit	Startup	Capacity (Mw)	Type of System	Sulfur Content of Coal (%)	Boiler Energy Consumed (%)	Makeup Water Consumed (gpm)	Design Removal Rates (%)		Average (%)		
								SO_2	Particulates	Availability	Reliability	Operability
Northern States Power	Sherborne 1	5/76	740	Alkaline Fly Ash	0.80	2.7	---	50	99	93.7	100	85.0
Northern States Power	Sherborne 2	4/77	740	Alkaline Fly Ash	0.80	2.7	---	50	99	95.2	100	88.7
Pacific Power & Light	Jim Bridger 2	2/80	550	Aqueous Sodium Carbonate	0.56	0.2	---	91	99	81.6	79.5	70.8
Pennsylvania Power	Bruce Mansfield 1	6/76	917	Lime	3.00	6.0	---	92.1	99.8	76.4	---	77.9
Pennsylvania Power	Bruce Mansfield 2	10/77	917	Lime	3.00	6.0	---	92.1	99.8	87.4	---	85.9
Pennsylvania Power	Bruce Mansfield 3	6/80	917	Lime	3.00	---	---	92.2	95	92.6	---	100
Public Service of New Mexico	San Juan 1	4/78	361	Wellman-Lord	0.80	4.4	365	90	99.8	92.2	93	88.2
Public Service of New Mexico	San Juan 2	8/78	350	Wellman-Lord	0.80	4.6	355	90	ESP	84.8	80.9	62.4
Public Service of New Mexico	San Juan 3	12/79	534	Wellman-Lord	0.80	3.6	---	90	99.5	85.1	78.1	62.8
Salt River Project	Coronado 1	12/79	280	Limestone	1.00	4.3	270	66	99.87	---	---	---
Salt River Project	Coronado 2	10/81	280	Limestone	1.00	4.3	270	66	99.87	---	---	---

Company	Unit	Startup	Capacity (MW)	Type of System	Sulfur Content of Coal (%)	Boiler Energy Consumed (%)	Makeup Water Consumed (gpm)	Design Removal Rates (%)		Average (%)		
								SO_2	Particulates	Availability	Reliability	Operability
South Carolina Public Service	Winyah 2	07/77	140	Limestone	1.70	1.1	100	45	99.4	96.7	97.3	97.5
South Carolina Public Service	Winyah 3	06/80	280	Limestone	1.70	2.1	-	90	ESP	43.4	55.1	55.2
Southern Indiana Gas and Electric	AB Brown 1	03/79	265	Dual Alkali	3.35	0.8	-	85	99.5	96.0	95.7	90.8
St. Joseph Zinc	GF Weaton 1	01/80	60	Citrate	2.00	-	-	90	99.6	-	-	-
Southern Mississippi Electric Power	RD Morrow 1	08/78	124	Limestone	1.30	5.5	-	52.7	99.6	48.3	62.2	71.2
Southern Mississippi Electric Power	RD Morrow 2	06/79	124	Limestone	1.30	5.5	-	52.7	99.6	37.2	45.6	70.1
Springfield City Utilities	Southwest 1	09/77	194	Limestone	3.50	4.6	-	80.0	99.7	41.6	35.0	33.5
Springfield Water, Light and Power	Dallman 3	01/81	205	Limestone	3.30	5.9	239	95.0	99.5	-	-	-
Texas Power and Light	Sandow 4	01/81	382	Limestone	1.60	-	-	75.0	99.7	-	-	-

Company	Unit	Startup	Capacity (MW)	Type of System	Sulfur Content of Coal (%)	Boiler Energy Consumed (%)	Makeup Water Consumed (gpm)	Design Removal Rates (%)		Average (%)		
								SO_2	Particulates	Availability	Reliability	Operability
Texas Utilities	Martin Lake 1	10/78	595	Limestone	0.90	1.3	550	71	99.4	-	-	-
Texas Utilities	Martin Lake 2	05/78	595	Limestone	0.90	1.3	550	71	99.4	-	-	-
Texas Utilities	Martin Lake 3	12/79	595	Limestone	0.90	1.3	-	71	99.4	-	-	-
Texas Utilities	Monticello 3	10/78	800	Limestone	1.50	-	546	74	99.5	-	-	-
Tennessee Valley Authority	Widow's Creek 8	01/78	550	Limestone	3.70	4.7	-	70	99.5	77.1	98.9	77.7
Utah Power and Light	Hunter 1	05/79	360	Lime	0.55	-	-	80	99.5	100.0	-	95.6
Utah Power and Light	Hunter 2	06/80	360	Lime	0.55	-	-	80	99.5	99.9	-	100.0
Utah Power and Light	Huntington 1	05/78	366	Lime	0.55	1.6	300	80	99.5	50.0	-	86.2

Chapter 3

FGD Systems and How They Work

This chapter describes the FGD systems available in the United States. The lime and limestone scrubbers, with which there is the greatest amount of operating experience, are taken up first, and the still experimental dry scrubbing systems are discussed last.

Scrubbers can be differentiated by three major characteristics:

1. whether the scrubbing process results in a wet or dry product;
2. whether the process uses a slurry or a solution to absorb sulfur dioxide, and;
3. whether it produces a saleable product or one that is thrown away.

However, these distinctions which are in widespread use, are not always used consistently.

Scrubbers also differ in:

1. the amount of sulfur dioxide and particulates that can be removed;
2. availability, reliability and operability;
3. the energy that is consumed, and;
4. the amount of water needed to make up for what is thrown out as waste.

These characteristics are considered by a utility when choosing one of four major families of scrubbers:

1. wet slurry throwaway scrubbing, in which the by-product is disposed of (including lime or limestone systems, magnesium enriched lime or limestone systems, lime or limestone systems with adipic acid, alkaline fly ash systems, and gypsum-producing systems);
2. wet solution throwaway scrubbing, in which the by-product is also disposed of (including the aqueous sodium carbonate, and the dual-alkali systems);
3. wet sulfur-producing scrubbing (including the magnesium oxide, and the Wellman Lord systems);
4. dry scrubbing, which produces a dry waste (including the spray drying and the dry injection systems [see Table A]).

As every power plant has to meet different emission standards, burns different kinds of coal, and employs a different method of waste disposal to meet environmental regulations, each scrubber is custom-built for its particular site.

Approximately 90 percent of the scrubbers in operation today are wet scrubbers. Most of these were designed to remove 85 percent of the sulfur dioxide from the flue gas, in order to comply with regulations promulgated under the 1971 New Source Performance Standards (NSPS). Those newer plants upon which construction began after September 18, 1978 must meet more stringent standards requiring as much as 90 percent removal, and therefore were designed to remove 95 percent of the sulfur dioxide from the flue gas in order to provide room for fluctuations in sulfur dioxide emissions.

Because of their high design sulfur dioxide removal capabilities, wet scrubbers are usually chosen by utilities burning a high-sulfur coal. However these systems also produce a wet muddy sludge as a waste.

Dry scrubbers, which are in an earlier stage of development, might be chosen by a utility burning low-sulfur coal and which must meet less stringent emission requirements or, if the utility is located in an area where there is not enough water available for a wet scrubber.

The choice of a reagent can have an important effect on the scrubbing process. If a reagent is water soluble, the scrubber is easier to operate and maintain than if the reagent is *not* water soluble. However, a water-soluble reagent also produces a water-soluble waste, which can

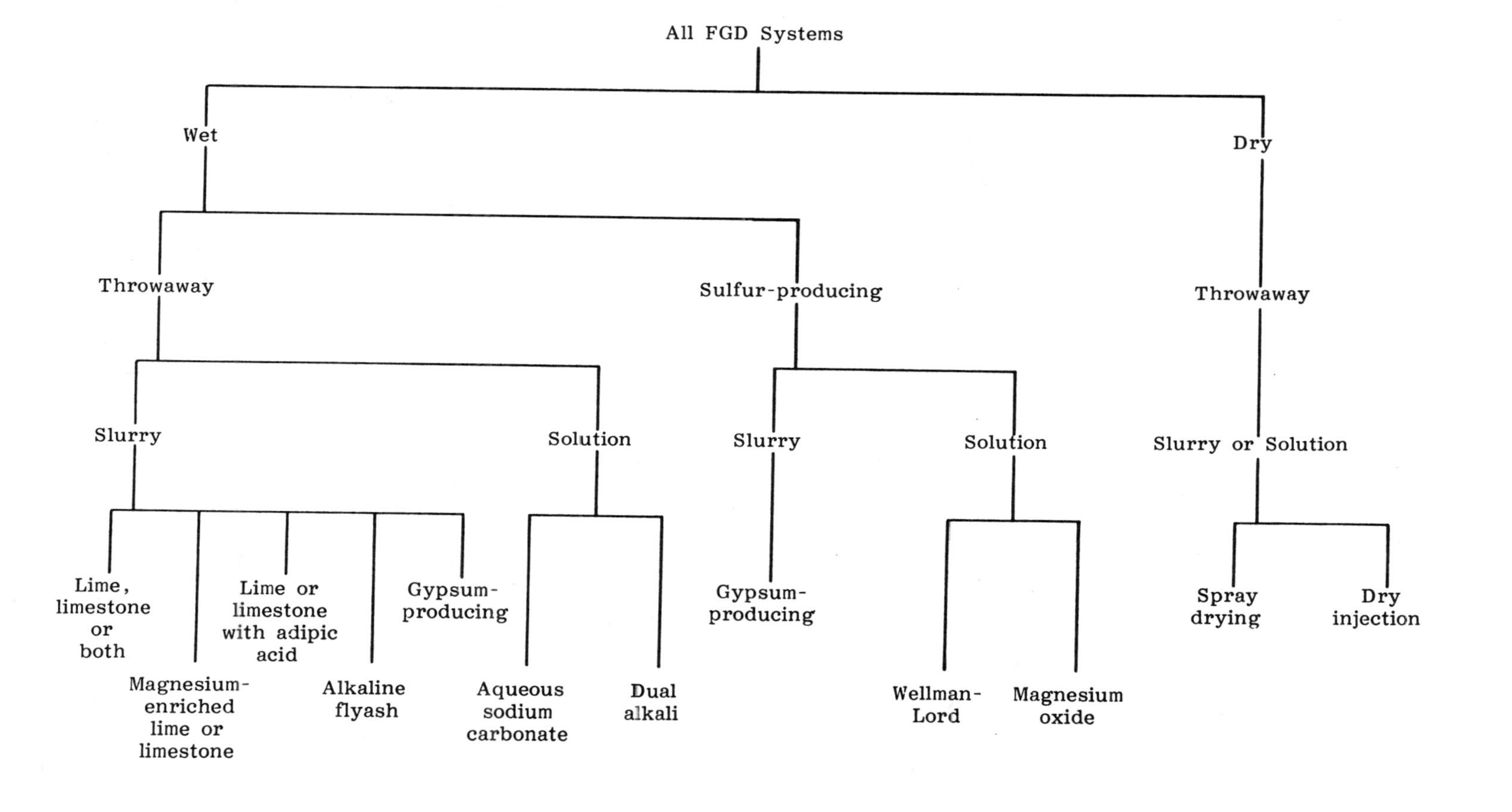
All FGD Systems
Wet
Dry
Throwaway
Sulfur-producing
Throwaway
Slurry
Solution
Slurry
Solution
Slurry or Solution
Lime, limestone or both
Magnesium-enriched lime or limestone
Lime or limestone with adipic acid
Alkaline flyash
Gypsum-producing
Aqueous sodium carbonate
Dual alkali
Gypsum-producing
Wellman-Lord
Magnesium oxide
Spray drying
Dry injection

Table I

TYPES OF FGD SYSTEMS IN OPERATION
(as of December 1980)

	Mw of boilers served by scrubbers	*Total Mw capacity served by scrubbers (%)*
Limestone	11,484	40.7
Lime	8,801	31.2
Alkaline flyash	4,093	14.5
Sodium carbonate	925	3.3
Dual alkali	1,201	4.3
Magnesium oxide	0	0
Wellman-Lord	1,540	5.5
Spray drying	110	0.4
		99.9 %

leach (or soak) into the ground water and pollute it, if it is not disposed of properly. The insoluble solids in slurry wastes pose less of a risk.

If the reagent is a sodium compound (e.g., sodium carbonate), it is water soluble; that is, it will form a solution when mixed with water. If the reagent is not completely water soluble, like limestone (calcium carbonate, $CaCO_3$), then the result of mixing it with water will be an absorbent slurry.

Some scrubber systems produce a sulfur, sulfuric acid or gypsum by-product which is potentially saleable. Processes that produce sulfur or sulfuric acid also produce some waste products, while processes that produce gypsum do not. The capital expenses and energy requirements of systems that produce saleable sulfur or sulfuric acid are considerably higher than the costs of other scrubbing systems as they include large chemical plants. However, systems that recover a sulfur product do not have a large waste disposal problem. In addition, the income from the sale of sulfur or sulfuric acid should help offset the high capital costs of these systems.

THE PROCESS OF ABSORPTION

The physical process is generally the same for most scrubbing systems. Absorption, the physical and chemical process whereby sulfur dioxide (SO_2) is oxidized to sulfite (SO_3) and sulfate (SO_4) molecules, takes place inside the absorber. The absorbent, a mixture of water and chemical reagent, will react with the sulfur dioxide as described above and remove it from the flue gas. The chemical reagent is usually an alkali such as calcium in the form of lime or limestone, or sodium in the form of dissolved minerals, e.g., nahcolite or trona.

The absorbent mixture is sprayed down from nozzles set into the sides and across the top of the absorber and strikes the flue gas which enters the bottom of the absorber and rises through it. Most of the sulfur dioxide in flue gas is absorbed by the droplets of absorbent in midair.

Once the absorbent liquid has come into contact with the flue gas, it remains in the absorber for only a few seconds. Used absorbent liquid drains continuously into an outside holding tank while fresh absorbent is pumped in to replace it. The used absorbent, now laden with reacted sulfur dioxide, remains in the holding tank while solids form. These solids and the used absorbent are then disposed of as waste. Alternatively, the used absorbent is pumped to a chemical plant where sulfur or sulfuric acid is produced. In the process, some of the absorbent is recovered and returned to the absorber where it is reused (see figure 1 for a simplified diagram of a wet scrubber system).

The flue gas, now cleaned of sulfur dioxide, continues to rise. Water droplets caught up in the rising flue gas are removed at the top of the absorber. The temperature of the flue gas drops as it passes through the absorber, so some systems must reheat the acidic gas to minimize condensation on the sides of the smokestack, and prevent corrosion of its liner. Finally, the treated gas is released to the atmosphere.

Wet Slurry Throwaway Scrubbing

Lime or Limestone Scrubbing

The most common methods of flue gas desulfurization in the United States are the lime and limestone wet scrubbing systems which, with variations, accounted for 86 percent of the scrubbers in use in December 1980.[1] Both lime and limestone scrubbers employ a process close to the one outlined above; the systems differ in the reagents used, and the performance obtained with each.

Limestone scrubbers have been designed to remove up to 95 percent of the SO_2 in flue gas; the median sulfur dioxide design removal rate of all operating limestone scrubbers is about 73 percent. Lime scrubbers

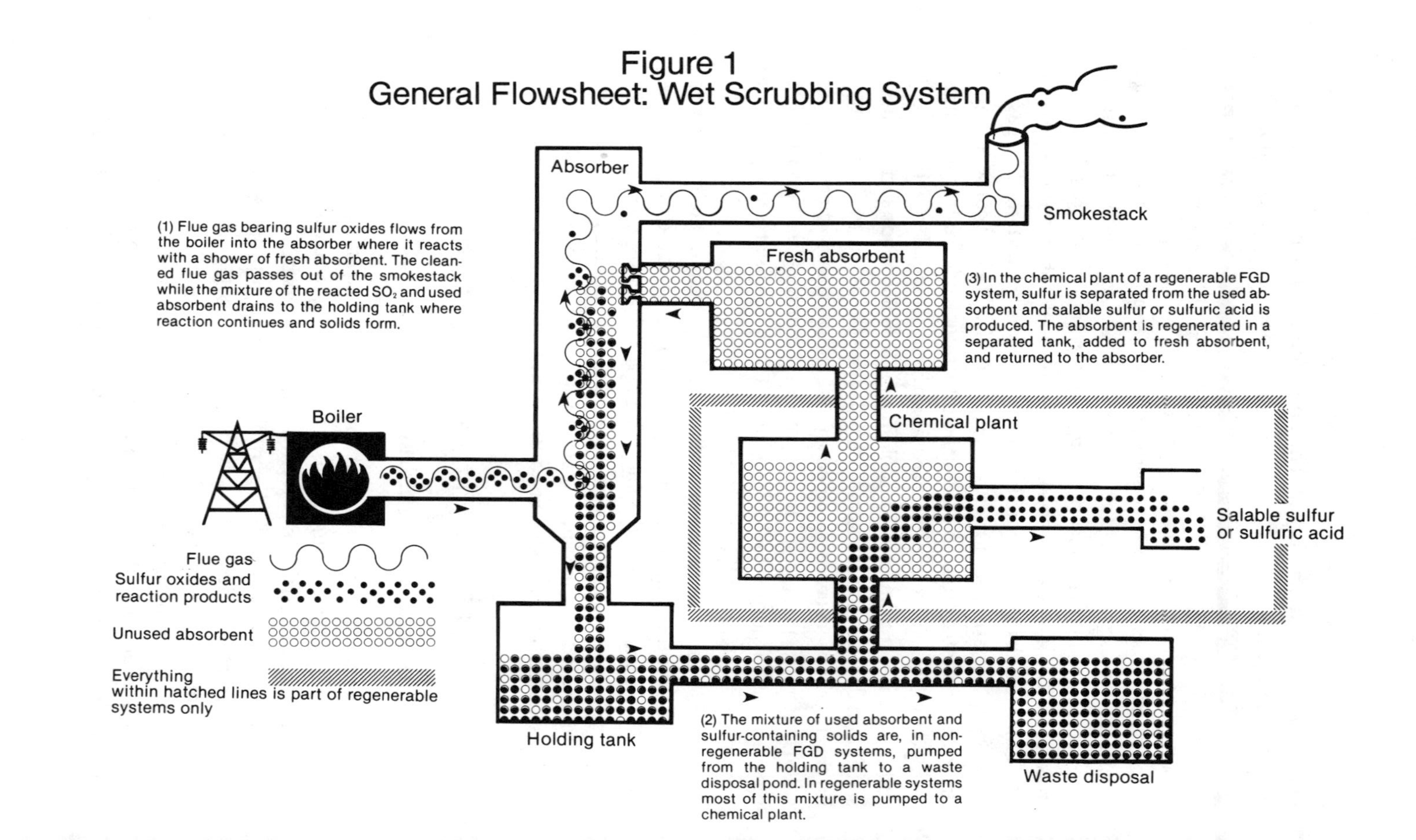
Figure 1
General Flowsheet: Wet Scrubbing System
Absorber
Smokestack
(1) Flue gas bearing sulfur oxides flows from the boiler into the absorber where it reacts with a shower of fresh absorbent. The cleaned flue gas passes out of the smokestack while the mixture of the reacted SO_2 and used absorbent drains to the holding tank where reaction continues and solids form.
Fresh absorbent
(3) In the chemical plant of a regenerable FGD system, sulfur is separated from the used absorbent and salable sulfur or sulfuric acid is produced. The absorbent is regenerated in a separated tank, added to fresh absorbent, and returned to the absorber.
Boiler
Chemical plant
Salable sulfur
or sulfuric acid
Flue gas
Sulfur oxides and
reaction products
Unused absorbent
Everything
within hatched lines is part of regenerable
systems only
Holding tank
(2) The mixture of used absorbent and sulfur-containing solids are, in non-regenerable FGD systems, pumped from the holding tank to a waste disposal pond. In regenerable systems most of this mixture is pumped to a chemical plant.
Waste disposal

have been designed to remove 92.2 percent of the sulfur dioxide. However, the median design sulfur dioxide removal rate of all lime scrubbers is 85 percent.

In general, the lime and limestone systems have lower capital, and operating and maintenance costs than all other available wet scrubbers. Their minimal equipment requirements keep the system's capital costs down, while their widely available reagents and low energy requirements (median electrical energy requirements are 3.25 percent of what is produced by the boiler for all limestone systems and 2.8 percent for lime) keep operating and maintenance costs lower than other systems. Because they were the first systems to be developed and the most widely used, there is a large body of operating experience leading to their improved design.

However, the best-known processes also have the best-known and most widespread problems. First, vast quantities of waste sludge, made up of calcium sulfate, calcium sulfite, unreacted lime or limestone and water are created, which must be impounded or treated to be used as landfill (see chapter on waste disposal). Second, the performance of these systems suffers if they are not operated with great care. The equipment can become plugged* with solids deposited in the scubber's nozzles, ducts, and drains, or coated with scale (the precipitation of very hard solids in thin layers on all of the surfaces of the scrubber system) and corroded. Third, these processes use their reagents only once, and not always efficiently, and there is, therefore, the continual cost of replacing the reagents.

Most other wet slurry scrubbers are modifications of these scrubbing systems. Four variations of the lime and limestone processes will be discussed below. One uses magnesium-enriched lime or limestone to enhance the absorption process. A second adds an organic acid, adipic acid, to improve the scrubber's operation. A third variation, the alkaline fly ash process, uses alkaline particulates from the ash as the reagent, together with additional lime or limestone to reduce costs. The final variation produces gypsum as a reaction by-product.

Magnesium-Enriched Lime and Limestone Scrubbing

Like calcium, magnesium can react with sulfur dioxide in the flue gas of a coal-fired boiler. The reaction products, magnesium sulfite

---

*Plugging can be largely prevented by good engineering design.

($MgSO_3$) and magnesium sulfate ($MgSO_4$), are more soluble than the analogous calcium reaction products. Hence the magnesium reaction products can accumulate to greater concentrations in the absorbing liquid without forming solids than can the calcium reaction products. Therefore, when the absorbent is made from lime or limestone that contains magnesium, it can absorb more sulfur dioxide without increasing the problem of scaling resulting from the precipitation of insoluble calcium products.

The process of lime or limestone scrubbing with magnesium is the same as the basic lime or limestone processes. The magnesium-enriched reagent is commercially available as magnesium-enriched lime or dolomitic limestone. Magnesium-enriched lime or limestone is slightly more expensive. For example, in 1980, PEDCo estimated that magnesium-enriched lime cost $61 a ton as compared to $46 a ton for regular lime.[2]

The waste, as in the lime or limestone processes, is disposed of in a lined or unlined pond, or a landfill. However, since waste magnesium products are more likely to pollute the groundwater than similar calcium compounds, the waste sludge must be treated or impounded.

Lime or Limestone Scrubbing with Adipic Acid

Adipic acid, a commercially available organic acid used in the nylon industry, and also as a food additive, has been tested as an additive to full-scale lime and limestone scrubbers for the past year. Adipic acid controls the acidity of the absorbent as the amount of sulfur dioxide entering the absorber fluctuates, and thus minimizes the formation of scale and improves the efficiency of the absorbent. (If the absorbent becomes too acidic, its sulfur dioxide absorption efficiency decreases.)

The addition of adipic acid to a limestone slurry has the added advantage of increasing the solubility of the limestone, so that less limestone is required and less waste is generated. There are however, some problems with this additive, as adipic acid tends to decompose, and some of the resulting products have a foul smell.

Lime Scrubbing

Number of scrubbers (as of December 1980)
In operation: 22
Under construction: 16
Total: 38

Megawatt Capacity of Boilers (as of December 1980)
In operation: 8,801
Under construction: 8,901
Total: 17,702

Characteristics of Lime Scrubbers

	Median	*Minimum*	*Maximum*
Design sulfur dioxide removal rate (%)	85	70	92.2
Design particulate removal rate (%)	99	0	99.8
Availability (%)	75.1 (mean)	0	100.0
Electrical energy consumption (%)	2.8	1.4	6
Make-up water (gpm)	150	50	700

Advantages, Disadvantages

- Advantage and disadvantages are the same as in limestone scrubbing.

Limestone Scrubbing

Number of Scrubbers (as of December 1980)
In operation: 34
Under construction: 48
Total: 82

Megawatt Capacity of Boilers (as of December 1980)
In operation: 11,484
Under construction: 25,366
Total: 36,850

Characteristics of Limestone Scrubbers

	Median	*Minimum*	*Maximum*
Design sulfur dioxide removal rate (%)	73 - 74	42.5	95
Design particulate removal rate (%)	99.5	0.0	99.87
Availability (%)	77.8 (mean)	34.2	100.0
Electrical energy consumption (%)	3.25	1.1	5.9
Make-up water (gpm*)	316	100	1,840

* gallons per minute

Advantages

- Lime or limestone scrubbing is a simple process with low capital and operating and maintenance costs.
- The process can achieve up to 90 percent sulfur dioxide removal. Additives such as adipic acid or the use of magnesium-enriched lime or limestone can make sulfur dioxide removal more efficient and reduce operating problems.
- The process has been in use since 1968, so a large body of operating experience has been built up.

Disadvantages

- The limestone process produces large quantities of waste sludge (about 1,500 tons a day for a 500 Mw plant) that require disposal in a way that is not harmful to the environment.
- The waste from lime or limestone scrubbing with magnesium-enriched lime or limestone can pollute groundwater.
- The scrubbing process has a tendency towards plugging, scaling and corrosion which can close down the scrubber. This will recur if the scrubber is not carefully designed and operated. It is difficult to limit the side reaction that produces gypsum.

Alkaline Fly Ash Scrubbing

The variation of the lime and limestone processes that uses alkaline fly ash in the absorbent is an imaginative application of scrubber technology. The absorption process is the same as in the lime and limestone systems. The difference is that instead of depending solely on outside supplies of calcium as the source of the absorbing alkali, boilers using low-sulfur western coal with a high alkaline content can use the alkaline matter in the fly ash, which would usually be a waste product, to make the absorbent slurry for the FGD system.

This is done by removing the fly ash, or particulates, from the flue gas with an electrostatic precipitator simultaneously with the sulfur dioxide; and throwing the fly ash into a holding tank with the absorbent slurries where the alkalis leach out of it. The slurry is then pumped to the absorber, where lime or limestone can be added as a supplemental alkali. Otherwise the scrubbing process proceeds in its usual manner.

The alkaline fly ash system has the advantage of requiring little equipment and therefore being easy to install. In addition, once it is dewatered, the waste sludge from this process (composed of fly ash and scrubber wastes) is more stable than the sludge from lime or limestone scrubbers, and can therefore be treated and used for landfill. However, because it does not achieve a very high level of sulfur dioxide removal unless lime or limestone is added, the alkaline fly ash scrubbing can only be used by utilities burning low-sulfur coals, and where local emission standards are less stringent.

For those utilities fortunate enough to have a source of low-sulfur coal, and that are located where emission standards allow them to take advantage of this process, substantial savings resulting from the elimination or reduction in reagent requirements may be realized. An alkaline fly ash system is in use at Minnesota Power & Light's Clay Boswell plant Unit 4 burns 0.94 percent sulfur coal and must meet a 1.2 lb. $SO_2/10^6$ Btu emission standard. The scrubber is designed to remove 84.55 percent of one sulfur dioxide, and in 1980, had an availability of 100 percent.

Alkaline Flyash Scrubbing

Number of Scrubbers (as of December 1980)
In operation: 11
Under construction: 2
Total: 13

Megawatt Capacity of Boilers (as of December 1980)
In operation: 4,093
Under construction: 1,400
Total: 5,493

Characteristics of Alkaline Flyash Scrubbers

	Median	*Minimum*	*Maximum*
Design sulfur dioxide removal rate (%)	60	36.13	84.55
Design particulate removal rate (%)	99.5	99.00	99.73
Availability (%)	87.4 (mean)	52.40	100.00
Electrical energy consumption	2.7	1.30	3.30
Make-up water (gpm)	535	370	895

Advantages

- The process is easy to retrofit, and cheaper than lime/limestone systems to build, operate and maintain.
- The waste sludge is more stable and less polluting than the waste from lime/limestone systems.

Disadvantages

- The system is limited to areas of the U.S. where power plants burn low-sulfur coal with a high alkali content.
- Particulates cause plugging problems in this system; erosion and corrosion are also operating problems.

Gypsum-producing systems

The gypsum-producing scrubbers are variations of the limestone systems. Calcium sulfite ($CaSO_3$), which is produced as a waste by lime and limestone scrubbers, is converted to gypsum ($CaSO_4$) by a process of forced oxidation, in a reaction that must usually be contained to prevent the plugging of scrubber equipment. The gypsum-producing scrubbers can therefore eliminate the problems of plugging and scaling, while achieving actual removal rates of 90 to 95 percent.

The potential market for this abatement gypsum is uncertain. Gypsum is used as wallboard by the construction industry, but in the United States naturally-occurring deposits are abundant, and cheaper than the abatement gypsum. However, gypsum is in demand in Japan where there are no natural reserves.

At present, only very minor amounts of abatement gypsum have been sold. Utility applications of this system will, for the most part, produce gypsum to throw away, although Louisville Gas & Electric Company is getting in position to sell abatement gypsum at its Trimble County plant, and other demonstration work on recovering gypsum is underway in this country. Despite the difficulties of marketing abatement gypsum, the other advantages of this system may well encourage its wider use.

The two Japanese companies that developed this process have designed new equipment for it. In the Chiyoda system, the entire scrubbing process takes place in one vessel. DOWA Mining Company's system uses a dual-alkali type process (see description below) with different reagents.

Wet Solution Throwayway Scrubbing

The seven wet scrubbers that use an absorbent solution have one main advantage over the scrubbers that use an absorbent slurry, in that the problems of plugging and scaling are controlled. Because these processes use a reagent that dissolves in water to form a solution, the reaction product is also water soluble. Hence no solids precipitate and form scale, and there are no free-floating solids to plug the equipment.

However, water soluble reagents are generally more expensive than insoluble reagents. PEDCo estimates the cost of limestone to be $12 a ton, and the cost of lime to be $50 to $60 a ton; in contrast, PEDCo estimates that sodium carbonate costs $150 a ton (in 1981 dollars).

Both the aqueous sodium carbonate and the dual-alkali systems described below form large quantities of potentially harmful wastes. Both systems, however, reduce the amount of this waste by recirculating some of their reagents through the absorber and reusing them.

Aqueous Sodium Carbonate Scrubbers

The four units in the United States using aqueous sodium carbonate scrubbers are all located in Nevada and Wyoming. (INFORM studied three of these units located at the Nevada Power Company [NPC]; see profile). This scrubbing process uses trona, a naturally occurring acidic sodium carbonate which can be used to absorb sulfur dioxide when dissolved in water. This process minimizes some of the operating and chemical problems that arise in slurry scrubbing systems.

This process proceeds much like the lime or limestone processes until the used absorbent reaches the holding tank, where the acid solution is neutralized with alkaline solids. Some absorbent liquid is drained from the system with the solid wastes and sent to a waste disposal system. Thus, some water and sodium carbonate must be added to the absorbent that is returned to the absorber to make up for what is lost as waste.

Aqueous sodium carbonate scrubbing is more readily applicable in western states because most naturally occurring deposits of trona are found there, and because one proven method of waste disposal used by Nevada Power Company is only usable in this region. NPC pumps its waste into clay-lined ponds where the water in the wastes evaporates, leaving a solid residue in the pond that is eventually carted away. (Clay-lined ponds prevent the waste from leaching into the groundwater.) It would be impossible to use this method of waste disposal in areas such as the eastern United States where rainfall and humidity are greater.

Aqueous Sodium Carbonate Scrubbing

Number of Scrubbers (as of December 1980)
In operation: 4
Under construction: 5
Total: 9

Megawatt Capacity of Boilers (as of December 1980)
In operation: 925
Under construction: 2,230
Total: 3,155

Characteristics of Aqueous Sodium Carbonate Scrubbers

	Median	*Minimum*	*Maximum*
Design sulfur dioxide removal rate (%)	90	85	91
Design particulate removal rate (%)	98	97	99
Availability (%)	84.6 (mean)	81.6	88.4
Electrical energy consumption (%)	5.0	0.2	5.0
Make-up water (gpm)	155	-	-

Advantages

- The system uses a reagent that is found in mineral deposits in large quantities in many areas of the U.S.
- The system is simple and minimizes chemical and equipment problems; the clear liquid absorbent, for example, minimizes plugging and scaling problems.

Disadvantages

- The waste sludge can cause serious pollution. Since large, lined waste-disposal areas are required, this could limit the process to low-humidity western areas.
- The non-regenerable process requires a steady supply of sodium carbonate. Though the reagent is available at present, the extent of its reserves could be limited.

Dual-Alkali Scrubbers

The dual-alkali process, which uses two separate alkalis, one in the absorber and the other in the holding tank, combines elements of the lime and limestone processes and the aqueous sodium carbonate systems, and gains many of the advantages of both. The absorber uses a solution of sodium hydroxide or sodium sulfite and water, which reacts with the sulfur dioxide in the flue gas to form soluble sodium sulfate. This drains to the holding tank where the second alkali, calcium, is added in the form of lime or limestone.* The calcium replaces the sodium in the absorbent reaction products to form calcium sulfite and calcium sulfite.

In this manner, two major problems are resolved: 1) plugging and scaling problems are lessened because the insoluble solids that cause them form outside the absorber where they can do little harm; and 2) the waste is less polluting than that of an aqueous sodium carbonate system because the calcium waste solids are less likely to dissolve and leach into ground-water than sodium solids.

In addition, the dual-alkali process has lower maintenance costs than lime or limestone scrubbing because there are fewer problems of plugging and scaling, and the process uses its reagents efficiently.

One problem with the dual-alkali process results from a side reaction which converts sulfites to sulfates, which can cause scaling throughout the system. This scaling is controlled by using a concentrated solution of absorbent that forces some by-product sodium sulfate out of solution with the waste product so there are no seeds for sulfate crystals to grow on. Normally, less than six percent of the sodium is lost in this way and must be replaced.[3] However, the dissolved sodium compounds which are discarded may cause pollution problems.

In November 1980, one supplier, FMC, was advertising a dual-alkali scrubber, and claiming that it reduced operating problems, lowered operating and maintenance costs, and could cost nearly 30 percent less than a lime or limestone scrubber.[4] As the benefits of this scrubber system become better known, it may come into wider use.

---

*Limestone has been used with less success in the United States.

Dual Alkali Scrubbing

Number of Scrubbers (as of December 1980)
In operation: 4
Under construction: 2
Total: 6

Megawatt Capacity of Boilers (as of December 1980)
In operation: 1,201
Under construction: 842
Total: 2,043

Characteristics of Dual-Alkali Scrubbers

	Median	*Minimum*	*Maximum*
Design sulfur dioxide removal rate (%)	90	85	95
Design particulate removal rate (%)	99.5	99	99.5
Availability (%)	80.4 (mean)	52.2	96.0
Electrical energy consumption (%)	1.0	0.8	2.4
Make-up water (gpm)	278	-	-

Advantages

- The system reduces the operating problems of plugging and scaling.
- The process has lower operating costs than lime/limestone systems because it is liable to fewer operating problems.

Disadvantages

- A side reaction that produces calcium sulfate has to be controlled by forcing some by-product sodium compounds out of solution. This produces two effects: 1) the waste sludge is made potentially more polluting, and 2) costly sodium carbonate must be added continuously.

Sulfur Producing Scrubbing Systems

Two processes, magnesium oxide and Wellman-Lord scrubbing, use a series of chemical reactions and complicated equipment to produce saleable sulfur or sulfuric acid and to regenerate the reagents. In general, systems that produce sulfur or sulfuric acid have higher capital costs than lime or limestone systems, since sulfur plants, requiring additional equipment and more expensive materials must be installed. Sulfur plants also increase the complexity of the chemical reactions involved in operating a scrubber, and require large amounts of energy. Their higher operating costs could be offset by the reduced costs of reagents, by savings resulting from not having to build a large waste disposal system, and possible by income generated from the sale of sulfur or sulfuric acid—if it were sold.

Magnesium Oxide Scrubbing

Magnesium oxide (MgO) scrubbing is a new FGD process which will not be in commercial use until 1982. This system uses a regenerable magnesium oxide costing $266/ton as a reagent.[5] The scrubber produces concentrated sulfur dioxide, which can be used to make sulfur or high-quality sulfuric acid. Moreover, the chemical plant that regenerates the magnesium oxide and produces sulfur or sulfuric acid does not have to be located immediately adjacent to the scrubber. One treatment plant can thus serve several scrubbers, making the MgO system easier to install and better suited for application in large cities where space is limited.

Advocates of this technology claim that this system has reduced plugging and scaling problems, as the reaction product is fairly soluble. It also recovers its expensive reagent, and eliminates disposal problems as there is very little waste.

However, the process does have some economic and operating drawbacks. Its energy requirements are high, and the generation of sulfur or sulfuric acid requires access to or construction of an expensive chemical plant. The availabilities of the prototypes were quite low (between 30 percent and 40 percent) because of operating problems such as corrosion and difficulty in handling waste magnesium solids. In 1981, only TVA and one other utility were building magnesium oxide scrubber systems (see Philadelphia Electric Company profile.)

Wellman-Lord Scrubbing Systems

The Wellman-Lord process has been adapted to utility use from oil-fired industrial boilers. Although this made utilities slightly more willing to accept it, the Wellman-Lord process had to be demonstrated on a full-sized scale before it was regarded as a proven technology. The first Wellman-Lord scrubbers for utilities were built by Northern Indiana Public Service in 1977 and Public Service of New Mexico in 1978. There are now seven Wellman-Lord scrubbers operating.

The Wellman-Lord process removes 90 percent of the sulfur dioxide in the flue gas and can be used with high- or low-sulfur coal anywhere in the United States. The process is complex, mostly because of the regeneration and sulfur-production stages. It uses an absorbent solution of sodium sulfite in water. Once the used absorbent is drained from the absorber, some is thrown out as waste, and the rest is heated to reverse the absorbing reaction and produce concentrated sulfur dioxide. The regenerated sulfur dioxide and sodium sulfite absorbent are separated. The gaseous sulfur dioxide is transported to a chemical plant to be made into pure sulfur of sulfuric acid. The sodium sulfite is redissolved in water, and sodium is added to it to replace what is lost as waste. The resulting absorbent liquid is then returned to the absorber.

Despite its complexity, the Wellman-Lord process has several advantages. Plugging and scaling problems are eliminated. It has been commercially proven by two utilities, and other utilities can also benefit from the experience of industrial users of the system. The high capital costs of the system (Public Service of New Mexico's four units' costs ranged from $126/Kw in 1978 (year of startup) to an estimated $295/Kw in 1982) could be partly offset by the sale of high-quality sulfur or sulfuric acid.

The Wellman-Lord process also has several problems. A side reaction that produces calcium sulfate, complicates the regeneration process and produces a solid that is difficult to remove from the system and dispose of as waste. If it is left in an unlined settling pond, the waste sodium produced can leach into the groundwater and pollute it. The process also uses an expensive absorbent inefficiently. Another drawback of the Wellman-Lord system is that the step reversing the reaction of the absorber requires large amounts of energy. In addition, the space requirements of the sulfur plant can make the system difficult to install or to use where space is limited.

Wellman-Lord Scrubbing

Number of Scrubbers (as of December 1980)
In operation: 7
Under construction: 1
Total: 8

Megawatt Capacity of Boilers (as of December 1980)
In operation: 1,540
Under construction: 534
Total: 2,074

Characteristics of Wellman-Lord Scrubbers

	Median	*Minimum*	*Maximum*
Design sulfur dioxide removal rate (%)	90	90	90
Design particulate removal rate (%)	99.5	0	99.8
Availability (%)	NA	NA	NA
Electrical energy consumption (%)	4.4 -4.6	3.6	12.1
Make-up water (gpm)	355-365	355	365

Advantages

- Operating problems such as plugging and scaling are virtually eliminated.
- The process produces much less waste than lime or limestone scrubbers.
- Utility operators can benefit from the experience of industrial operators of the system.
- The system produces high quality sulfur dioxide that can be converted to pure sulfur or sulfuric acid.
- The process can be used with any high or low-sulfur coal anywhere in the U.S.

Disadvantages

- The process can require a very.high percentage, up to 12%, of the energy produced by the plant it serves.
- The process produces a solid waste that must be disposed of with care to prevent it from polluting.

Dry Scrubbing

"Dry scrubbing" is any scrubbing process that produces a waste containing less than 5 percent water. Of these, the throwaway processes are receiving the most attention. Two such processes, spray drying and dry injection, are in development stages today. Lower requirements for emissions reductions in the western U.S. have contributed to the development of the spray dryer.

Generally, the primary potential advantages of dry systems over wet systems are: 1) lowered capital, operation and maintenance costs resulting from simpler design; 2) reduced waste disposal problems, since there are no wet sludges to be handled and disposed of; and 3) significantly less maintenance, since the absence of wet absorption products, or waste sludge, in the absorber eliminates the plugging and scaling problems that are the bane of an FGD operator's existence.

While they look promising, dry scrubbers have not yet been commercially demonstrated. The first system to come into operation, at Northern States Power's Riverside 6-7 station, in November 1980 is designed to remove 90 percent of the sulfur dioxide.

Dry scrubbers also have their own operating problems, as described below.

Spray Drying

"Spray drying" is a vivid description of the physical events that take place during this process. The flue gas in a spray dryer is at a temperature of 250° to 400° F. As it flows through the absorber, the gas passes through a fine mist of dissolved or partly-dissolved alkalis. The amount of moisture added is controlled so that the flue gas is only partially saturated and the reaction products remain dry. As in the wet process, the mist can be a solution or a slurry. The alkalis react with the sulfur dioxide and are suspended in the gas stream. Since the solids are dry, they will not plug the absorber or form scale. The same is true of the fly ash. The gas stream carries all of these solids to a particulate collector (a baghouse or an electrostatic precipitator) where the solids are removed from the gas stream and disposed of as waste. The cleaned flue gas is then released to the atmosphere (see figure 2).

The choice of a reagent is important to the design of a spray dryer scrubber as it is for wet scrubbers; although this system does not incur the problem of unplugging equipment fouled by insoluble solids in the waste products. Sodium carbonate (Na_2CO_3), lime and trona ($Na_2CO_3 \cdot NaHCO_3 \cdot H_2O$), (soda ash) have all been tested as reagents. However, the differences among reagents must be considered by utilities building dry scrubbers. The choice depends on the particular needs of the utility. If a utility burns coal with a medium (2 percent) sulfur content, it may choose one of the sodium-based alkalis, which are by far the most reactive reagents, removing 80 to 90 percent of the incoming sulfur dioxide. Because of their molecular structure they are also a cheaper reagent. Lime is a more practical choice for utilities using such coal. Lime used once removes 45 to 60 percent of the sulfur dioxide; if it is recycled and used again, the percentage increases to 80 to 85 percent. Limestone is not acceptable as a reagent because it removes less than 30 percent of the sulfur dioxide.

Another consideration in choosing a reagent is the waste product produced. Dry scrubbing's main benefit derives from the fact that its dry waste product is easy to handle. However, if the waste is a sodium-based product from a sodium-based reagent, it is water soluble and can pollute the groundwater if it is dissolved by rainfall. Therefore it is used in arid regions where evaporation significantly exceeds rainfall. In addition, dust control can be a problem. Consequently, the dry waste must be managed in a way that prevents these problems.

The utility's choice of a particulate collector for the spray drying process (which may be a baghouse or electrostatic precitator (ESP)) also affects sulfur dioxide rates and energy costs. Unlike an ESP, a baghouse provides additional surface area where sulfur dioxide absorption can take place, because unreacted alkalis on the surface of the bag can react with incoming sulfur dioxide. On the other hand, an electrostatic precipitator is less sensitive to condensation but it provides no surface for the further removal of sulfur dioxide.

Various studies by the TVA Office of Power have estimated that the capital costs of a lime spray dryer FGD system ranged from $80/kw in 1980, depending on the location of the plant and the sulfur content of the coal. The capital costs for a trona (soda ash) spray dryer on a boiler burning low-sulfur western coal were estimated to range from $152/kw to $160/kw in 1980.

Spray Drying

Number of Scrubbers (as of December 1980)
In operation: 1
Under Construction: 10
Total: 11

Megawatt Capacity of Coilers (as of December 1980)
In operation: 110
Under Construction: 3,413
Total: 3,523

Characteristics of Spray-Drying Scrubbers

	Median	*Minimum*	*Maximum*
Design sulfur dioxide removal rate (%)	90	-	-
Design particulate removal rate (%)	99.5	-	-
Availability (%)	NA	-	-
Electrical energy consumption (%)	NA	-	-
Make-up water (gpm)	NA	-	-

Advantages

- The dry waste product is easy to handle.
- The system has minimal operating problems such as plugging and scaling.
- The system should be less expensive than wet scrubbing systems to install, operate and maintain.

Disadvantages

- The dry waste product is a potential pollutant if it is dissolved in water. Controlling dust is also a problem.
- The system may be limited to power plants burning low-sulfur coal.

Dry Injection

When the dry injection is used, the need for the absorber, already reduced in importance in the spray drying system, disappears altogether. The dry absorbent, usually nahcolite, a form of sodium bicarbonate, is forced into a stream of flue gas and onto the filter of a baghouse. The absorbing reaction takes place midstream; as in a spray dryer, the product is caught up in the flue gas. The gas goes to a baghouse where the resulting product and fly ash are removed and the cleaned gas is released to the atmosphere. Most of the sulfur dioxide removal seems to take place on the filter of the baghouse. The reagent is added to the system through a duct leading to the baghouse, in the baghouse itself, or in both of these places.

Most dry injection systems tested use nahcolite as a reagent because it is very reactive. Large reserves of this mineral exist in the U.S., so it is potentially very cheap. However, in 1981 it was still not mined commercially. Moreover, waste products from dry injection systems that use nahcolite as a reagent cause the same problem as wastes from spray dryers using the reagent: the sodium-based waste is a potential pollutant. The possibility of using other reagents such as trona is being researched.

Despite these problems, the nahcolite injection process seems to be a viable, if undeveloped, option for utilities planning to build scrubbers that are located in areas where evaporation significantly exceeds rainfall. Its equipment requirements and maintenance cost are minimal, and the EPA estimates the capital costs of a nahcolite injection system to be $50/kw in 1980. However, by 1981 the process was not developed enough to begin planning any scrubbers in the near future; and if the preferred reagent, nahcolite, remains commercially unavailable and no substitute is found, the prospects of this process may remain questionable.

Figure 2
Flowsheet: Spray Drying Scrubbing System

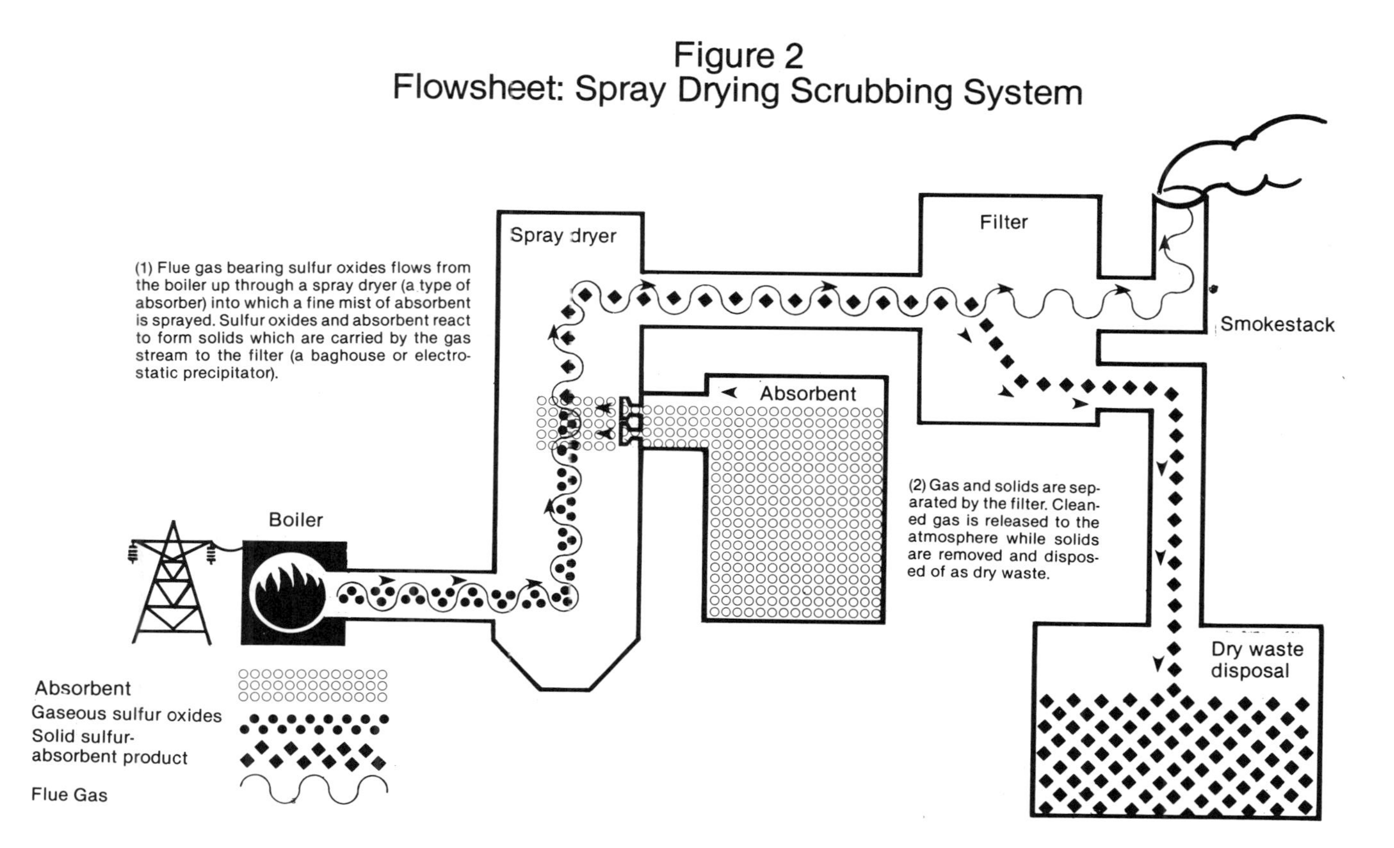

Chapter 4

Treatment and Disposal of FGD Wastes

Flue gas desulfurization systems reduce the concentration of sulfur dioxide expelled into the air, but, in the absorbent process wet lime and limestone systems transform chemicals, minerals and unexpelled parts of this sulfur pollutant into a solid waste. Most of the 84 FGD systems operating in 1980 were of this type and they produced significant quantities of waste. Projections have been made that some 23 million tons will be generated by such scrubbers in the year 1985.[1] A scrubber serving a 1,000 Mw coal-fired power plant burning a coal with 3.5 percent sulfur and 14 percent ash can produce 14.4 million tons of dry FGD waste and fly ash over a 30-year period. This is enough waste to cover approximately 540 acres with 30 feet of sludge.[2] Finding suitable locations for disposal of such quantities of sludge poses difficult problems, but they are not insurmountable.

Regulations and Hazards

The disposal of FGD waste is regulated by many laws, the most important of which is the Resource Conservation and Recovery Act (RCRA) of 1976. RCRA directs EPA to develop and enforce regulations that govern the disposal of hazardous and nonhazardous wastes in ponds, landfills and mines.

Tests conducted by Arthur D. Little, Inc. for EPA on a number of FGD waste samples have indicated that trace contaminants are present in concentrations that may be considered toxic.[3] EPA, however, has classified FGD waste as nonhazardous pending the completion of this study. If this study indicates that the liquid portion of FGD waste contains more than 100 times the concentration of any of the 21 contaminants listed in the National Interim Primary Drinking Water Standards, then such waste may be considered hazardous. Preliminary investigations have shown that the concentration of dissolved pollutants in FGD sludges varies enormously, with some wastes exceeding the concentration limits set by the interim water standards, and others not.

Until these studies are completed, FGD wastes must be disposed of in a manner that complies with RCRA's nonhazardous waste criteria as outlined by EPA. These criteria specify that endangered species, surface waters, groundwater and wetlands should not be adversely affected by the disposal of solid wastes. Each state is responsible for developing and enforcing its own solid waste management plan to meet the nonhazardous waste criteria.

First Step: Treating The Slurry

The slurry that drains from the holding tanks of wet lime or limestone FGD systems has the consistency of a soup containing 5 to 15 percent solids. In this form it is highly unstable and difficult to handle. The exact composition of this mixture varies widely, depending on the particular FGD system, the boiler, the amount of sulfur dioxide removed, the composition of the coal before burning and whether or not the waste is mixed with dry fly ash before disposal.

The solid portion of the alkaline waste produced by wet lime and limestone scrubbers is composed primarily of calcium sulfate, calcium sulfite and unreacted lime or limestone. The liquid portion contains dissolved substances such as sulfate, chloride, fluoride, lead, mercury, iron, selenium and manganese that could pollute drinking water sources (see Table I). The disposal of FGD sludge containing such contaminants is no simple matter.

Since slurry from wet lime and limestone FGD systems contains only 5 to 15 percent solids, an enormous amount of water can be discarded with the waste. To salvage some of this water and reduce the volume of waste, most utilities use a "thickener" to cause the solids in the slurry

Table I

Contents of the FGD Waste Produced from Lime and Limestone FGD Systems

Solid Portion	*Liquid Portion*
calcium sulfate ($CaSO_4$)	sodium
calcium sulfite ($CaSO_3$)	calcium
calcium oxide (lime)	magnesium
calcium carbonate (limestone)	chlorine
	sulfate
	sulfite

*Trace Contaminants**

antimony	lead
arsenic	manganese
barium	mercury
beryllium	molybdenum
boron	nickel
cadmium	selenium
chromium	silver
cobalt	tin
copper	uranium
flourine	zinc
iron	

* in both solid and liquid portions

Source: Michael Baker, Jr., *FGD Sludge Disposal Manual*, (Palo Alto, CA: Electric Power Research Institute, January, 1979).

to settle. The layer of clear water that forms on top of the solids is decanted and reused in the scrubbing process. The remaining waste, or sludge, which has a solids concentration of 25 to 40 percent, is drained from the thickener and usually pumped directly to a disposal pond.

Disposal in Ponds

The disposal pond acts as a settling basin where the particles in the sludge again sink to the bottom, leaving a layer of water above. Water pollution can result from the disposal of untreated FGD wastes in a pond if groundwater is below the pond and no barrier or liner exists between the pond and the soil beneath it. In these situations, the weight of the water pushes contaminated liquid through the sludge and underlying soil into the groundwater. This liquid, or leachate, seeps through different types of soil at different rates. Sandy soil, for example, allows water to drain through it quite rapidly while a clay soil acts as a barrier that prevents such drainage into the groundwater.

The pollution of groundwater can be minimized by lining the pond or providing a drainage system under the pond. Liners made of clay and synthetic materials have been used to trap the contaminated liquids within the disposal site, and a number of utilities, such as Central Illinois Light, Kentucky Utilities, and Alabama Electric Cooperative, have lined their sludge disposal ponds. Still, the long-term desirability of such liners has not yet been determined.

Another alternative, the installation of an underdrainage system, consists of a series of perforated pipes that collect the leachate as it percolates through the waste and return it to the scrubber. This system has been successfully demonstrated at TVA's Shawnee Power Plant in Paducah, Kentucky.

Disposal in Landfills

There has been a trend in recent years to dispose of FGD wastes in landfills rather than ponds. In order for FGD sludge to be used as landfill, the solids content must be increased and the waste stabilized.

Vacuum filtration and centrifugation can transform FGD sludge into a semisolid waste. Vacuum filtration, as its name implies, uses vacuum suction to separate the solid and liquid portions of the FGD sludge; a filter collects the solids as the water is sucked through it. In centrifugation, the sludge is spun around at high speeds until distinct solid and liquid layers form. Both of these "dewatering" methods produce a waste having a solids content that ranges from 50 percent to as high as 80 percent for sludge from gypsum-producing scrubbers.

Dewatered FGD waste, however, is physically unstable unless treated further. It cannot be disposed of as landfill because it might reliquify when disturbed or shaken, and because it can absorb rainwater and again become sludge.

Chemical Treatment

To minimize the risks of groundwater contamination, and to insure its stability as landfill, dewatered FGD waste must be chemically treated or fixated. Chemical fixation with fly ash and lime, for example, hardens the FGD sludge to a clay-like material. Also it can reduce by some 1,000 percent the waste's permeability by water, as compared to that of untreated sludge. Moreover, landfills composed of such clay-like waste are stable enough to be used in the future as sites for parks, cemeteries, and light structures such as warehouses.

The chemical fixation methods currently available include alkaline fly ash blending, Calcilox and Poz-O-Tec. Alkaline fly ash blending consists of mixing fly ash that has a high lime content (up to 36 percent) with FGD sludge. The waste can be disposed of in a pond or landfill, depending upon the amount of water removed from the waste. The Calcilox process, developed by the Dravo Lime Company, uses a proprietary additive and magnesium enriched lime to produce a stable waste that can be disposed of in a pond or landfill. Pennsylvania Power's Bruce Mansfield Plant for example, disposes of its waste through Calcilox ponding (see profile). The most popular method of chemical fixation, however, is the Poz-O-Tec process developed by I.U. Conversion Systems. It uses fly ash and lime to stabilize dewatered FGD waste for disposal in a landfill. Duquesne Light Company and Louisville Gas & Electric Company are now operating such a waste disposal system with success (see profiles).

An improperly designed landfill can nullify some of the benefits derived from chemical fixation. To avoid this, certain engineering practices can be followed including: 1) compacting the waste to further reduce its permeability; 2) grading the site and providing a good drainage system to handle runoff; and 3) reclaiming the site with topsoil and vegetation to prevent wind or water from eroding its surface.

Although chemically-fixated FGD waste is usually disposed of as landfill, innovative utilities have been using their FGD waste as a base material to pave roads, parking lots and airports. The Duquesne Light Company, for example, has used some of the chemically-treated waste from its Elrama Station to pave an ash-haul road at its Cheswick Station. These applicaions, however, are not expected to consume a significant portion of the FGD waste that is generated.[4]

Mine Disposal

In addition to putting wastes into settling ponds and landfills, coal-burning power plants may dispose of their FGD wastes by dumping them into the unused portions of nearby surface (strip) and underground coal mines. Mine disposal of chemically treated wastes has apparent advantages, helping to restore the original contour of strip-mined land, minimizing ground subsidence of "cave-ins," over underground mines, and improving the quality of the water near the mine by neutralizing the acidic drainage in the mine.

This disposal method has a significant drawback, however: sulfur dioxide gas is generated by the reaction between the acid drainage in the mine and the FGD waste. The amount of sulfur dioxide that can be released has not yet been determined. In 1980 only six power plants were using surface coal mines as disposal sites for their detwatered but otherwise untreated FGD waste.[5]

Ocean Disposal

Power plants located in the northeastern United States that are expected to convert from oil to coal generally have limited amounts of land available for the disposal of FGD sludge and fly ash. Primarily for this reason, experiments were conducted between 1976 and 1980 by a team of scientists at the State University of New York at Stony Brook to determine the feasibility and environmental impact of building artificial ocean reefs from chemically stabilized blocks of FGD sludge, fly ash and lime. Preliminary tests in the lab and ocean indicated that such blocks were not harmful to marine life. In fact, those blocks placed in the ocean soon became overgrown with seaweeds and marine life.

Encouraged by these results, the research staff built, in September 1980, an artificial reef composed of 18,000 blocks along three miles of the south shore of New York's Long Island. This $2.9 million experiment, scheduled to continue for three years, is being funded jointly by the Federal EPA, the U.S. Department of Energy, the Electric Power Research Institute, the New York State Energy Research & Development Authority, and the Power Authority of the State of New York.[6,7] Environmentalists and government agencies alike hope that such artificial ocean reefs will not only provide sites for utility waste disposal, but will promote local fisheries as well.

Disposal Costs

The costs of disposing of FGD waste and fly ash can amount to as much as 25 percent of the capital investment for an entire scrubbing system. As of mid-1979, these costs ranged from $30 to $48 per Kw produced by a power plant, according to a study conducted by the Tennessee Valley Authority.[8] As shown in Table B, the disposal of chemically treated waste in a landfill has the lowest capital costs, while the disposal of the same waste in a pond has the highest.

Table II

THE COST OF FGD WASTE AND FLYASH DISPOSAL

Disposal options	*Capital costs in $/Kw (mid-1979)*	*Annual expenditures in mills/Kw (mid-1980)*
Untreated waste in a clay lined pond	34.4	0.94
Sludge flyash blending in a landfill	36.4	1.64
Chemical fixation in a pond	48.2	1.91
Chemical fixation in a landfill	30.4	1.76
Surface mine disposal of flyash stabilized waste	35.3	1.54

Basis: 500 Mw power plant, 18-year life, 7,000 hours per year revenue requirement basis; 3.5% sulfur, 16% ash coal; limestone FGD system

Source: Veitch, J.D., Steele, A.E., and Tarkington, T.W., 1980, *Economics of Disposal of Lime/Limestone Scrubbing Wastes: Surface Mine Disposal and Dravo Landfill Processes*, U.S. Environmental Protection Agency, EPA-600/7-80-022, p. xxiii.

In this same study by TVA the annual operating expenditures for FGD waste disposal ranged from 0.9 to 1.9 mills per Kwh in mid 1980 (see Table II). Pond disposal of untreated sludge has the lowest annual operating cost, while pond disposal of chemically treated waste, again, has the highest cost, making the latter the most expensive option (see Pennsylvania Power's profile).

Chapter 5

Profiles

FGD Supplying Firms

AIR CORRECTION DIVISION OF UOP, INC.
101 Merritt - 7
P.O. Box 5440
Norwalk, CT 06856

Goods and Services	Air Correction Division of UOP designs, develops and installs a variety of FGD systems. Its other air pollution control equipment includes: electrostatic precipitators, flyash conditioning systems, fabric filters and centrifugal collectors.
FGD Systems	Utility contracts obtained as of May 1981 Operating: 15 (3746 Mw) Planned or under construction: 1 (720 Mw)
Parent Company	The Signal Companies, Inc. 9665 Wilshire Boulevard Beverly Hills, California 90212 Revenues (1980): $4,383,500,000 Operating Profit (1980): $167,700,000

The Air Correction Division of UOP is one of three FGD suppliers in the United States that offers four sulfur dioxide scrubbing processes: lime, limestone, sodium carbonate and DOWA dual-alkali. Thus, a utility can choose the system that will best meet its site-specific requirements. From 1970 to May 1981, ACD obtained 16 contracts for utility FGD systems, 15 of which are now operating (see Table A).

The DOWA dual-alkali process is the only one offered by Air Correction Division which has not yet been used on a full-scale, coal-fired boiler. This dual-alkali process is unique in that it uses limestone instead of lime as the regeneration chemical, and it produces a high quality gypsum as a by-product. Since the market for such gypsum is limited, Air Correction Division projects that utilities will sell only 10 percent of the gypsum produced by the DOWA process to the wallboard and cement industries, and dispose of the remaining 90 percent.

The DOWA process was originally developed for use on oil-fired boilers by the DOWA Mining Company of Tokyo, Japan. In 1977 UOP obtained the exclusive license to sell and manufacture the DOWA process in the United States, and it plans to extend this license to Europe.

Research and development of flue gas desulfurization technologies at Air Correction Division has been directed toward: (1) refining its lime, limestone, and sodium systems; (2) demonstrating and refining its DOWA dual-alkali process at Tennessee Valley Authority's Shawnee Plant in Paducah, Kentucky; and (3) examining the physical and chemical characteristics of the waste produced from limestone and DOWA scrubbers.

Air Correction Division is also developing a new process using copper oxide as the absorption reagent. The Company is working on this project in cooperation with Shell Oil Company at Tampa Electric's Big Bend Station. The process is distinctive in that it can remove both sulfur dioxide and nitrogen oxides from the flue gas. However, Air Correction Division says that it is too expensive to be used by utilities as long as the control of nitrogen oxides is not required by the federal government.

AMERICAN AIR FILTER CO., INC.

215 Central Avenue
Louisville, Kentucky 40277

Goods and Services	American Air Filter supplies lime FGD systems, electrostatic precipitators, fabric-filter dust collectors, and heating, cooling and ventilating systems.

FGD Systems — Utility contracts obtained as of May 1981

Operating:	6 (1,428 Mw)
Planned or under construction:	2 (810 Mw)

Parent Company

Allis-Chalmers Corporation
Box 512
Milwaukee, Wisconsin

Revenues (1980): $2,063,940,000

Net income (1980): $48,222,000

American Air Filter is one of the smaller FGD supplying firms, having obtained eight contracts for lime or limestone systems since 1973, all but two of which have been built. American Air Filter also offers dual-alkali and gypsum-producing systems, but as of May 1981 these systems had not yet been sold. Research and development projects on the production of commercial-grade gypsum and the disposal of FGD wastes are being conducted within American Air Filter's Engineering Department. Most of the funding for these and other research projects is generated in-house.

BABCOCK & WILCOX

Fossil Power and Construction
20 S. Van Buren Avenue
Barberton, Ohio 44203

Goods and services

Babcock & Wilcox (B&W) is a supplier of both wet and dry FGD system. B&W also produces steam-generating systems for fossil-fuel and nuclear power plants, computerized control systems for power plants, boiler-cleaning equipment electrostatic precipitators, baghouses, air heaters, fans, steel tubing and machine tools.

FGD systems — Utility Contracts obtained as of May 1981

Operating:	9 (4070 Mw)
Planned or under construction:	10,(4,552 Mw)

Parent company

J. Ray McDermott & Co., Inc.
1010 Common Street
New Orleans, Louisiana 70122

Revenues (1980)	$3,282,510,000
Net income (1980)	$ 88,366,000

Babcock & Wilcox is the third largest supplier of FGD systems in the United States, with contracts for 17 wet lime and limestone scrubbers and two dry scrubbers (see Table A). Babcock & Wilcox is one of the two suppliers interviewed that manufactures many of the components used in its FGD systems, such as electrostatic precipitators, baghouses and fans.

Babcock & Wilcox began to research dry scrubbing in 1977. By 1980 it had obtained contracts from Basin Electric Power Cooperative and Colorado Ute Electric Association for large-scale, lime/spray drying units. These will be installed on boilers which burn low-sulfur (0.54 percent and 0.70 percent, respectively) western coal. Babcock & Wilcox's research on dry FGD systems is directed toward adapting this process for use with higher-sulfur coals, increasing the efficiency with which the lime reagent is used, and developing useful applications for the dry waste produced.

COMBUSTION ENGINEERING, INC.

Environmental Systems Division
Post Office Box 43030
31 Inverness Center Parkway
Birmingham, Alabama 35243

Goods and services

Combustion Engineering (CE) is an FGD supplier that specializes in lime- and limestone-based systems. This highly diversified energy company is the largest supplier of fossil-fuel electrical-generating systems, and the third largest supplier of nuclear reactor components, in the U. S. CE also sells electrostatic precipitators, fabric filters, reheaters, ash-handling systems, resource-recovery systems, oil and gas drilling equipment, gas-processing plants, pulverizers, rotary dryers, and various glass products, including solar collector glass. In addition, CE is involved in a coal-gasification program, synfuels and fluidized-bed technology.

FGD systems	Utility contracts obtained as of May 1981	
	Operating:	16 (6,410 Mw)
	Planned or under construction:	14 (8,958 Mw)

Combustion Engineering is the leading supplier of FGD systems in the United States; it has sold 30 units for utility boilers having an electrical capacity of 15,368 Mw (see Table A). These systems are lime, limestone, lime/alkaline fly ash, limestone/alkaline fly ash, and forced-oxidation/gypsum by-product processes. Combustion Engineering has an advantage over most other FGD suppliers, as it can manufacture many of the components that are used in its FGD systems.

Combustion Engineering, one of the first companies to explore this technology, began researching flue gas desulfurization in 1964. In 1968, two years before the Clean Air Act was passed, Combustion Engineering installed a limestone FGD unit on a 125 Mw boiler at the Lawrence power plant owned by Kansas Power and Light. Combustion Engineering and Kansas Power and Light have been working together since then to improve the design of Combustion Engineering's FGD systems, and to examine both the disposal and the marketability of gypsum produced from Combustion Engineering's scrubbers.

Combustion Engineering has developed a spray-drying FGD system which removes particulates via a baghouse produced by the James Howden Hilma Company of the Netherlands. Combustion Engineering has not yet obtained a contract for this technology, but it is optimistic about future sales.

GENERAL ELECTRIC ENVIRONMENTAL SERVICES, INC.
200 North 7th Street
Lebanon, Pennsylvania 17042

Goods and services

General Electric Environmental Services designs and manufacturers both wet and dry FGD syste It also provides electrostatic precipitators, fabri filters, mechanical collectors, industrial wet scrubbers, and waste treatment and disposal systems. This company was formed in April 19 when General Electric brought two divisions of the Envirotech Corporation, Chemico Air Polluti Control of New York and Buell Emission Control of Lebanon, Pennsylvania.

FGD Systems

Utility contracts obtained as of July 1981

Operating:	5845 Mw (22)
Planned/ Under Construction:	5394 Mw (10)

Parent Company

General Electric Company
1 River Road
Schenectady, New York 12345

Revenues (1980):	$5,714,000,000
Net Income (1980):	$ 407,000,000

General Electric Environmental Services is the second largest supplier of FGD systems in the United States, with 32 contracts for scrubbers that service boilers having a total electrical capacity of 11, 239 Mw. This company was formed in April 1981 through the acquisition of the Chemico and Buell divisions of Envirotech Corporation by the General Electric Company.

General Electric Environmental Services offers the following FGD processes: lime, limestone, force oxidation/gypsum by-product, lime/alkaline fly ash, dual alkali, magnesium oxide and lime/spray drying. Its FGD technology has been exported to Japan, Holland and Germany.

This company also offers sludge treatment and disposal systems such as fly ash/sludge bending, and chemical fixation with lime and fly ash. (See waste disposal section).

NIRO ATOMIZER INC.

Oakland Ridge Industrial Center
9165 Ramsey Road
Columbia, Maryland 21045

Goods and services

Niro Atomizer is a leading worldwide supplier of spray dryers, an important component of dry FGD systems. Niro also produces fluid-bed dryers, flash dryers, evaporators, and closed-cycle dryers. Its equipment has been used by the dairy industry, in the production of coffee, baby food and instant starch; by the chemical industry, in the production of detergents, fertilizers and pesticides; by the mineral industry, in processing copper, iron and lead; by the pharmaceutical industry, in the production of aspirin, vitamins, and penicillins; and by the polymer industry, in the production of poly vinylchloride (PVC).

FGD systems

Utility contracts obtained as of May 1981

Operating:	1 (110 Mw)
Planned or under construction:	6 (2,242 Mw)

Parent company

A/S Niro Atomizer
305, Gladsaxevej
DK-2860 Soeborg, Denmark

Niro Atomizer has been researching dry FGD technology since 1974 at its headquarters in Copenhagen, Denmark. In 1977 Niro entered into an agreement with the Western Precipitation Division of Joy Manufacturing, located in Los Angeles, to design and market a lime-based dry scrubber. As of May 1981 Niro and Joy have obtained seven FGD contracts from utilities for lime/spray dryers, six of which have not yet been built. They recently completed the first large-scale utility application of spray-drying technology, which began operation at Northern States

Power's Riverside Station in Minneapolis in January 1981. The operating data from this 100 Mw demonstration system is being scrutinized by utilities, FGD suppliers and government agencies alike to ascertain how well dry scrubbers work. Niro's joint venture with Joy Manufacturing will extend until 1984.

At its research facility in Copenhagen, Niro Atomizer is examining the ability of fly ash to remove sulfur dioxide, the mechanism of the sulfur dioxide absorption reaction, and the properties, possible use, and disposal of dry FGD waste.

PEABODY PROCESS SYSTEMS
835 Hope Street
Stamford, Connecticut 06907

Goods and services

Peabody Process Systems designs, engineering a installs limestone and lime/alkaline flyash FGD systems. It also provides particulate-removal equipment such as electrostatic precipitators, fabric filters, and scrubbers.

FGD systems

Utility Contracts obtained as of May 1981

Operating:	6 (2,085 Mw)
Planned or under construction:	5 (2,875 Mw)

Parent company

Peabody International Corporation
4 Landmark Square
Stamford, Connecticut 06901

Revenues (1980) :	$399,276,000
Net Loss (1980) :	$ (8,075,000)

Since the early 1970s, Peabody Process System has obtained 11 utility contracts for limestone and lime/alkaline fly ash FGD systems, five of which have not yet been built. Peabody's FGD sales have increased quite rapidly since the successful completion of its contract with Alabama Electric Cooperative in 1979 and with Minnesota Power and Light in 1980.

Peabody feels that its sulfur dioxide scrubbers are unique because: (1) they have sophisticated control systems—with a one-button start-up feature—that allow them to be operated by a minimum number of personnel; (2) they consume very little energy, only 0.8 to 1.3 percent of a boiler's electrical capacity; (3) they use the sulfur dioxide absorbing reagents very efficiently; and (4) they can be bought by utilities for a price of $75 per Kw (1981 dollars), which includes every element of the scrubber from gas inlet to gas outlet to dewatering equipment.

RESEARCH-COTTRELL, INC.

P.O.Box 1500
Somerville, New Jersey 08876

Goods and services	Research-Cottrell supplies both wet and dry FGD systems. It also designs fabric filters, constructs tall chimneys and cooling towers, engineering wastewater-treatment facilities, and produces equipment for the drilling and distribution of oil and gas.
FGD systems	Utility contracts obtained as of May 1981
	Operating: 11,4,872 Mw)
	Planned or under construction: 5 (2,014Mw)
Revenues (1980)	$379,594,000
Net Income (1980	$ 7,535,000

Research-Cottrell is the fourth largest FGD supplier in the United States, having sold 16 systems to utilities for boilers with a total electrical capacity of 6,886 Mw. These systems include a limestone process that produces either waste or gypsum, and a lime/spray drying process. Research-Cottrell has engaged the services of Knauf, a large gypsum supplier in West Germany, to produce from Research-Cottrell's limestone scrubber a high-quality gypsum that can be used in the construction industry. Research-Cottrell's lime/spray drying process was developed through cooperation between this company and Komline-Sanderson: Research-Cottrell supplies the fabric filter; Komline-Sanderson supplies the spray dryer.

Research-Cottrell's FGD systems can be supplied complete with waste treatment and disposal equipment. Three FGD units have been sold which can achieve sludge stabilization by the addition of fly ash to form a semisolid material.

Research-Cottrell's corporate philosophy places a strong emphasis on research and development. 80 percent of the R & D funds for FGD are generated within the company, while 10 percent are obtained from government sources and 10 percent from utilities. The goals of this R & D program are to produce by-product gypsum that is suitable for use as wallboard, and to apply Research-Cottrell's dry FGD system to boilers that burn coals with a sulfur content greater than 1.5 percent.

THYSSEN-CEA ENVIRONMENTAL SERVICES, INC.

550 Madison Avenue
New York, New York 10022

Goods and services

Thyssen-CEA Environmental Systems designs and constructs FGD systems for utilities. It was formed in December 1980, when Combustion Equipment Associates sold its Air Pollution Control Group to Thyssen AG of West Germany.

FGD systems

Utility contracts obtained as of May 1981

Operating:	8(1,854 Mw)
Planned or under construction:	1(295 Mw)

Parent company	Thyssen AG Tostfach 67 4100 Duisburg West Germany

Thyssen-CEA Environmental Systems (formerly known as the Air Pollution Control Group of Combustion Equipment Associates) is one of the oldest FGD supplying firms, having sold its first scrubber in 1971. Since that time, it has obtained eight more contracts from utilities for FGD units, seven of which are now operating. Thyssen-CEA offers a number of different FGD processes: lime, lime/alkaline fly ash, sodium carbonate and dual-alkali. It is now developing a limestone process and a dual-alkali process which uses limestone instead of lime.

WHEELABRATOR FRYE INC.
600 Grant Street
Pittsburgh, PA. 15219

Goods and services

Wheelabrator-Frye and its subsidiary, the M.W. Kellog Co., supply dry and wet FGD systems, respectively. Wheelabrator-Frye also designs and manufacturers electrostatic precipitators, fabric filters, fans, cranes, conveyors, residential water meters, water/wastewater treatment and processing equipment, refuse-to energy systems, pigments, varnishes, carbon paper and "carbonless reproduction paper." The M. W. Kellog Company designs and constructs gas-processing plants, oil refinery units, fertilizer plants, and synthetic-fuel plants.

FGD Systems

Utility contracts obtained as of May 1981

Operating:	4 (2,057 Mw)
Planned or under construction:	4 (2,200 Mw)

Revenues (1980): $1,299,673,000

Net Income (1980):	$ 55,300,000
Subsidiary	M. W. Kellogg Company Northeast Operations Center 433 Hackensack Avenue Hackensack, New Jersey 07601

Before its acquisition of the M. W. Kellogg Company in September 1980, Wheelabrator-Frye supplied only dry FGD systems. Wheelabrator-Frye and Rockwell International Corporation had a joint venture contract to develop a technology for dry scrubbing, in which Wheelabrator-Frye contributed to the fabric-filter expertise, while Rockwell International contributed its spray-dryer expertise. Wheelabrator-Frye and Rockwell International obtained a contract from the Otter Tail Power Company in 1978 to equip a 410 Mw boiler at the Coyote Station with an aqueous carbonate/spray drying unit. (This dry scrubber began operating in May of 1981 and is designed to remove 70 percent of the SO_2 from the flue gas of a boiler that burns 0.87 percent sulfur coal.) In 1980 the association between Wheelabrator-Frye and Rockwell International ended, and each has entered the FGD market independently. However, as of May 1981 Wheelabrator-Frye had not obtained another contract for its dry FGD system.

According to Wheelabrator-Frye, the development and refinement of dry FGD technology, and the enthusiasm within the utility industry over its use, are a result of the New Source Performance Standards that require the removal of 70 percent of the SO_2 from the flue gas of boilers that burn low-sulfur coal. Wheelabrator-Frye is conducting research on a variety of dry injection processes using a number of different absorbents.

The M. W. Kellogg Company, a subsidiary of Wheelabrator-Frye, since September 1980, began its research on wet FGD systems in the late 1960s. Kellogg developed a sulfur dioxide absorption process based on the use of magnesium-enriched lime or limestone. The addition of magnesium sulfate serves to increase the absorption of sulfur dioxide from the flue gas and to significantly decrease the calcium sulfate and calcium sulfite scaling within the FGD unit. This process is known as the Kellogg-Weir process, because the horizontal spray absorber used was developed by Dr. Alexander Weir of Southern California Edison. Kellogg first sold this system in 1976 to the Pennsylvania Power Company for use on its Bruce Mansfield Plant. Since that time, six more scrubbers have been sold. Kellogg also offers an FGD system that produces a gypsum by-product that is suitable for use in the construction industry.

M. W. Kellogg's research and development goal is to design improved FGD systems that can be marketed at a cost that is comparatively low but consistent with trouble-free operation. The major R&D operations on FGD include enhancing sulfur dioxide removal and investigating the use of alternate reagents and construction materials.

Utilities

DUQUESNE LIGHT COMPANY
435 Sixth Avenue
Pittsburgh, PA 15219

Service Area:	Allegheny and Beaver Counties in southwestern Pennsylvania, including Pittsburgh.
Regulatory Agency:	Pennsylvania Public Utility Commission
Electricity Customers:	
Residential	500,466
Commercial	48,306
Industrial	2,005
Streetlighting and Other	1,725
Total	552,502
Total Electrical Capacity (1980):	3,199 Mw
Peak Load (1980):	2,474 Mw
Reserve Capacity:	29%
Source of Power:	89% coal 8% nuclear 2% oil 1% outside purchases
Total Operating Revenues (1980):	$689,465,000
Affiliated Electrical Group:	Central Area Power Coordination Group (CAPCO)

Source: 1980 Annual Report

Duquesne Light Company owns and operates three coal-burning power plants near Pittsburgh, Pennsylvania: Phillips, Elrama and Cheswick. Phillips and Elrama are older plants that have been retrofitted with FGD systems. The Cheswick plant was built in 1970 and uses low-sulfur coal (1.4 to 1.6 percent sulfur) without scrubbers.

In 1969 Duquesne Light commissioned its architect/engineering firm, Gibbs & Hill, Inc., to examine the available options for meeting the stringent sulfur dioxide and particulate standards that were soon to be promulgated by the State of Pennsylvania and Allegheny County. Flue gas desulfurization, low-sulfur coal, and low-sulfur oil were all studied as possible alternatives to comply with emission standards of 0.6 pounds of SO_2 per million Btu and 0.08 pounds of particulates per million Btu. The report, completed in 1970, concluded that flue gas desulfurization technology was the "most feasible means of sulfur dioxide control..."* However, the decision to use this technology was made at a time when there was virtually no commercial operating experience with scrubbers.

Installation and Operation

Chemico / Envirotech (now General Electric Environmental Services) obtained the contract to supply lime desulfurization systems for the Phillips and Elrama Stations in 1971. It took two years for the first phase of a two-phase FGD system to be fitted to the 29-year-old Phillips plant, and it was not until 1977 that the entire 387 Mw system was completed, at a cost of $137 per Kw.

The operation of this system was plagued with problems such as the chemical corrosion and physical breakdown. The corrosion and physical breakdown problems were minimized by coating the vulnerable equipment surfaces with a special material which is resistant to the acidic and highly abrasive substances that were damaging the FGD equipment. Scaling was controlled by cleaning the scrubber every 1,400 hours.

The FGD system at the 487 Mw Elrama station was completed in October 1975 at a cost of $125 per Kw. Because of the experience gained at the Phillips Station, Duquesne had fewer installation and start-up problems at Elrama.

---

*Steve L. Pernick, Jr. and R. Gordon Knight, "Duquesne Light Company, Phillips Power Station, Lime Scrubbing Facility," (Paper presented at The APCA 68th Annual Meeting, Boston, MA, June 15-20, 1975).

Duquesne Light has refined the operation of its FGD systems considerably. For example, the use of lime containing 8 to 10 percent magnesium oxide significantly reduced the scaling problems and significantly increased the reactivity of the lime absorbent. As a result of these and other improvements, the FGD systems at the Phillips and Elrama Stations were available for operation 79.7 percent of the time and 99.9 percent of the time respectively in 1980.

The Phillips and Elrama power plants are in compliance with the sulfur dioxide removal standard, which requires 83 percent sulfur dioxide removal using 2 percent sulfur coal, 80 percent of the time. As of October 1980, they have never been fined for failing to meet these requirements. However, a spokesman for the company asserted that it is very difficult, if not impossible, to meet the emissions standard (0.6 lb $SO_2/10^6$ Btu) continuously. This is because fluctuations in SO_2 emissions are related to the daily fluctuations in the performance of the scrubber and the sulfur content of the coal.

Waste Disposal

Duquesne Light's FGD systems and particulate collectors at the Elrama and Phillips Stations produce a total of 2,500 tons of dewatered waste per day. This waste is chemically stabilized by the Poz-O-Tec process, which involves mixing the FGD sludge with fly ash and lime, and dewatering the mixture to a solids content of about 70 percent (see chapter on waste disposal). The treated waste is then trucked to an area where it is used as landfill in a disposal site that is lined with both fly ash and clay, and has underground and surface drains to minimize groundwater contamination. A disposal site of 38 acres is used near the Elrama Station, and another 25 acres has been designated for this use near the Phillips Station; these sites will be filled to capacity in 1987 and 1990, respectively. In 1980 the disposal of FGD waste and ash in this manner cost Duquesne Light $11.3 million.

Table A

DUQUESNE LIGHT COMPANY
FGD SYSTEMS DATA

	Elrama Units 1 - 4	Phillips Units 1 - 6
Gross Capacity	510 Mw	408 Mw
Percent Utility Ownership	100	100

Commercial Start-up Date: Plant	1952 - 1961	1942 - 1956
Commercial Start-up Date: FGD System	October 1975	July 1973
Architect/ Engineering Firm	Gibbs & Hill, Inc.	Gibbs & Hill, Inc.
FGD Supplier	Chemico/Envirotech	Chemico/Envirotech
Type of FGD System	Lime	Lime
SO_2 Removal Rate (%)		
Design	83	83
Actual	86	87
Particulate Removal Rate (%)	99	99.3
SO_2 Emission Standard	0.6 lb./MBtu	0.6 lb./MBtu
Particulate Emission Standard	0.10 lb./MBtu	0.08 lb./MBtu
Coal Characteristics (%)		
Sulfur	2.0 to 2.3	2.0 to 2.3
Ash	15	16
Btu/lb.	11,500	11,800
FGD Availability (1980)	99.9	79.7
FGD Energy Consumption		
percent (%)	2.35	2.54
Mw	12	10
FGD Water Consumption (gallons/minute)	700	600
FGD Reagent Consumption (tons/day)	150 -190	100 - 120
Number of Persons Needed to Maintain and Operate FGD System	31	31

Costs

Duquesne Light has made a significant investment of time and money in controlling sulfur dioxide emissions from its coal-fired generating stations. After some initial difficulties, changing in operational procedures and equipment have significantly improved the performance of its FGD systems. However, these scrubbers are particularly expensive; in 1980, the annual cost of the FGD system at Phillips and Elrama was $44.7 million.

Table B

DUQUESNE LIGHT COMPANY
1979 COSTS

	Elrama Units 1 -4	Phillips Units 1 - 6
Capital Cost: Plant (includes scrubbers)	$158,000,000	$133,000,000
Capital Cost: FGD System $ (1980) $/Kwh	 $62,500,000 $125	 $53,000,000 $137
Annual Costs of FGD $/year mills/Kwh (FPC)	 $23,900,000 5.168	 $20,837,000 6.651
Annual Revenue Requirements: Plant	$82,300,000 (1980) $102,67/Kw	$70,800,000 (1980) $126.36/Kw
Annual Revenue Requirements: FGD System (1980)	$23,900,000 $8.36 mills/Kwh	$20,837,000 $10.7 mills/Kwh
Waste Disposal Costs FGD Waste $/year (1980)	 1.799 mills/Kwh $5,800,000	 2.655 mills/Kwh $5,500,000
Financing	County Pollution Control Bonds	

KANSAS CITY POWER & LIGHT
1330 Baltimore Avenue
Kansas City, MO 64141

Service Area:	4,700 square miles in western Missouri and eastern Kansas; population 1,000,000.
Regulatory Agencies:	Public Service Commission of Missouri; Public Service Commisssion of Kansas; Federal Energy Regulatory Commission; Nuclear Regulatory Commission
Electricity Customers:	
Residential	301,417
Commercial and Industrial	41,199
Other	148
Total	342,764
Total Electrical Capacity:	2,838 Mw
Peak Load:	2,198 Mw
Reserve Capacity:	29.1 %
Sources of Power:	Five steam electric generating stations, all coal fired. Gas and oil fired plants for peak loads. Coal, 93%; gas, 5.6%; oil, 0.9%.
Coal Source:	Local coal, Missouri and Kansas strip mines; Wyoming and other western mines under long-term contracts
Average Revenue Per Customer:	
Residential	6.023¢ per Kwh
Commercial	5¢ per Kwh
Industrial	4¢ per Kwh
Other	4.49¢ per Kwh
Total Operating Revenues (1980):	$440,182,000

Source: 1980 Annual Report

Five coal-fired steam electric generating stations provide 93 percent of Kansas City Power & Light's (KCP&L) electricity. KCP&L was among the first utilities in the U.S. to build a scrubber, and its experience has helped solve the operating problems of later scrubbers, and contributed a vast amount to the body of technical and operating knowledge of FGD technology. The newer of KCP&L's five stations, Iatan, which began operating in 1980 (too late for this study), and La Cygne, which began operating in 1973, are subject to sulfur emissions controls and have wet limestone scrubber systems.

KCP&L also retrofitted wet lime scrubbers on its Hawthorne Units 3 and 4. These systems, built by Combustion Engineering, use marble-bed absorbers which are no longer available. The utility provided very little additional information on the Hawthorne scrubbing systems (see tables).

The Choice of an FGD System

KCP&L decided to use a scrubber on an 820 Mw coal-burning boiler at its La Cygne station in the early 1970s because it wanted to use cheap local coal. This coal has a high sulfur content, (5 to 6 percent), so a scrubber was required to meet local sulfur dioxide emissions standards.

According to the *Wall Street Journal* in 1977, the utility had decided to build La Cygne's scrubber despite the high cost of the scrubber equipment because "KCP&L still felt it would save money in the long run by using the 'dirty' local coal," adding that "Today. . . the utility is convinced it made the right choice."

However by 1981 opinion within the utility appeared to be divided on this point. In a telephone interview, Wayne R. Johnson, Assistant to the President and Chief Environmental Engineer, stated that "If I had to put in a new unit tomorrow, I would do anything to avoid a scrubber." He said KCP&L spends about $38 million a year for all its pollution control efforts, which he seems to think is too much.

The new wet limestone scrubber (see Process chapter) that KCP&L built with La Cygne Unit 1 was designed to remove 80 percent of the SO_2 and 98.75 percent of the particulates from the flue gas. In 1975 a short full-boiler load (800 Mw to 830 Mw) test and a longer continuous capacity (700 Mw to 720 Mw) test showed 76.2 percent and 80.14 percent sulfur dioxide removal, respectively. Particulate removal for the second test was 98.2 percent.

After the construction of an additional scrubber module, the system was reported to have exceeded its design specifications.

Construction and Operation

KCP&L's wet limestone scrubber on La Cygne Unit 1 was built by

Babcock and Wilcox (B&W; see profile). KCP&L chose B&W over several other suppliers because B&W had supplied boilers to the utility. KCP&L did not use an architect/engineering firm and the utility's operations and maintenance personnel did not participate in the decisions concerning the scrubber.

The utility chose its limestone scrubber on the basis of price and the reliability of the equipment. The utility decided against sulfur-producing systems such as the Wellman-Lord scrubber because marketing studies found no market for sulfur or sulfuric acid. No performance guarantees were offered, as no suppliers offered them at the time the scrubber was built.

KCP&L's La Cygne Unit 1 scrubber did not work when it first started operating in 1973 because of numerous mechanical and chemical problems. The 1977 *Wall Street Journal* article described the early operating problems: "During the first summer [the scrubber] leaked, and during the winter it froze. Pipes and valves corroded, pumps failed, and fans wore out. Heavy deposits of a cement-like scale would collect inside the steamy scrubber chamber and shut it down. The scale buildup got so bad at times that workers would have to use sledgehammers to remove it."

In order to protect its investment, the utility poured money and manpower into the scrubber operation to make it work. The utility spent an extra $7 million, about $8/kw. Most of this money paid for the installation of an additional scrubber module. The rest went to devise and pay for new maintenance procedures.

KCP&L has greatly improved the unit's operating record through improved operating procedures and a better understanding of scrubber chemistry. In 1979 the La Cygne Unit 1 was working reliably about 95 percent of the time; this figure increased to 96 percent in 1980.*

The key to the success of KCP&L's program is a very high level of maintenance. To control scale formation and corrosion, each scrubber is cleaned and washed at least once a week. The spraying of the interior surfaces of the scrubber modules has eased the plugging problems by clearing away accumulated soft solids. Furthermore, many scrubber parts and interior surfaces were re-covered with harder materials such as stainless steel, which are less likely to be damaged by the slurry's acidity and which discourage scale from adhering. Most important, the maintenance team learned enough about the scrubber chemistry to control the slurry's acidity, which regulates the formation of scale.

---

*As reported in *EPA Utility FGD Survey: October-December 1980 Vol. 2*, EPA-600/7-81-012, Jan. 1981. Prepared by PEDCo. Environmental, Inc.

Table A

KANSAS CITY POWER & LIGHT
FGD SYSTEMS DATA

	LaCygne 1	Hawthorn 3	Hawthorn 4
Percent Utility Ownership	50	50	100
Capacity (Mw)	820	85	85
Initial Startup			
Boiler	1973	1951	1951
FGD System	Feb., 1973	Nov., 1972	Aug., 1972
Type of FGD System	Wet limestone	Wet lime	Wet lime
Architect/ Engineer	None	None	None
FGD System Supplier	Babcock & Wilcox	CEA (now Thyssen-CEA)	CEA (now Thyssen-CEA)
Design SO_2 Removal Rate	80%	70%	70%
Actual	75 -80%	70%	70%
Design Particulate Removal Rate	98.75%	98.6%	98.6%
Emission Standards			
SO_2	1.5 lb./MBt u	6.1 lb./MBtu*	
Particulate	0.128 lb./MBtu	0.15 lb./MBtu	
Coal Characteristics			
Sulfur	5 -6%	2.5%	2.5%**
Btu/lb.	9,000 - 9,700	11,317	11,317
Energy Consumption	2.7%	2.2%	2.2%
Reagent Requirement	500 tons of limestone per year	NA	NA
Water Requirement	1,148 gpm	NA	NA
Availability (1980)	89.5%	79.3%	88%

* Station limit set for the combined emissions of five Hawthorne boilers.

** A blend of high sulfur (10.6%) Oklahoma coal and low sulfur (0.8%) Wyoming coal.

Costs

The cost of KCP&L's achievement with La Cygne Unit 1 has been high. The utility had to make a $7 million capital expenditure on top of its original investment of $46.8 million ($68/kw). The operating and maintenance team requires 54 people, including 8 operators per shift as well as maintenance and administrative personnel. The *Wall Street Journal* reported in 1977 that "The company says it saves 30% [on coal costs] by using local coal instead of shipping coal from Wyoming 'without degrading the environment.' It would convert more units, the utility says, if enough local coal were available."

Table B

KANSAS CITY POWER & LIGHT
1979 COSTS

	La Cygne 1	Hawthorn 3	Hawthorn 4
Capital Cost of Boiler (in 1,000's)	$166,586	$15,200	$13,600
Per Kw	$208	$179	$160
Capital Cost of FGD System (in 1,000's)	$46,800	$3,255	$3,255
·Per Kw (FPC)	$68	$29 combined	
Annual Revenue Requirements of Boiler	NA	NA	NA
Annual Costs of FGD ($/year)	6,500	1,000 combined	
mills/Kwh (FPC)	3.469	2.754 combined	
Waste Disposal Costs	NA	NA	NA
Financing	NA	NA	.NA

The Future of FGD

Wayne Johnson cited three specific areas of FGD technology that are in need of further research. First, because the metals required for the actual construction of the scrubber are not commonly available, he thinks that research into the metallurgy of scrubber materials should continue along with research into scrubber chemistry. Second, alternative ways of reheating the flue gas need to be considered and developed. Third, the effects of particulates and particulate removal on scrubber performance need to be evaluated. The utility's research priority however, is the solution of day-to-day operating problems rather than the further examination of scrubber technology.

LOUISVILLE GAS & ELECTRIC COMPANY

311 West Chestnut Street
P.O. Box 32010
Louisville, KY 40232

Service Area:	Louisville and the surrounding area in Kentucky; 700 square miles; population 805,000.
Regulatory Angencies:	Public Service Commission of Kentucky; Federal Energy Regulatory Commission
Electricity Customers:	295,066
Total Electrical Capacity:	2,472 Mw
Peak Load:	1815 Mw
Reserve Capacity:	36.2%
Sources of Electric Generation:	97% Coal 3% Hydroelectric
Rates (average price):	4.109¢ per Kwh
Total Operating Revenues (1980):	$443,894,000

Source: 1980 Annual Report

Louisville Gas and Electric Company (LG&E) has received environmental awards from the New York Botanical Gardens, the Kentucky Department of Natural Resources and Environmental Protection, the National Society of Professional Engineers, and the Louisville Audobon Society for its role in the advancement of FGD technology, and for the reduction of air pollution from its coal-burning power plants. The Environmental Protection Agency (EPA) has worked closely with LG&E, contributing $6.5 million for the research and testing of scrubbing and FGD waste-disposal systems. Through this joint relationship, LG&E has gained considerable FGD expertise, which allows the company to play a major role in the planning and installation of its scrubbers. In fact, LG&E does its own engineering and construction work on its FGD systems, thus reducing capital costs of these systems.

Installation and Operation

LG&E's three coal-burning power plants, Paddy's Run, Cane Run and Mill Creek, are located in Jefferson County, Kentucky. As early as 1973, a 72 Mw boiler at the Paddy's Run Station was retrofitted with a demonstration lime FGD system. The scrubber enabled the boiler to meet the emission standard of 1.2 pounds of sulfur dioxide per million Btu (1.2 lb $SO_2/10^6$ Btu), while burning 2.5 percent sulfur coal. As a result of this demonstration, a plan was developed in 1975 by LG&E, the EPA and the Air Pollution Control District of Jefferson County requiring that eight of LG&E's 16 coal-burning boilers comply with the sulfur dioxide emission standard by installing FGD systems. The exceptions allowed by this plan are: (1) the first three units of the Cane Run plant, because they will be retired in 1985, and (2) five units at the Paddy's Run plant, because they are used only during periods of peak electrical demand.

As of April 1981, four full-scale lime scrubbers were in operation—on Cane Run Units 4 and 5 and Mill Creek Units 1 and 3. The operating costs of these lime systems are unusually low because they can use carbide lime, a waste product from a nearby acetylene plant, as the sulfur dioxide absorbent. This carbide lime is available at a very attractive price of $15 per ton (1980 dollars), as compared to $46 per ton (1980 dollars) for conventional lime.

LG&E's boilers with these FGD systems are capable of meeting the sulfur dioxide emission standard while burning a high-sulfur (3.75 percent) coal. In 1980 the scrubbers at Cane Run 4 and 5 and Mill Creek 3 were available for operation 94.7 percent, 82.7 percent and 66.7 percent of the time, respectively; similar data from Mill Creek Unit 1 have not yet been compiled. The systems with low availabilities were plagued by mechanical failures in equipment such as pumps and dampers. According to Robert Van Ness, manager of Environmental Affairs:

> "We've got the [FGD] chemistry down; that's no problem; but when you have to rely on the performance of mechanical equipment that isn't directly related to the scrubbing process, that's where the problems begin."

However, in spite of these mechanical problems, all of LG&E's lime systems had an average SO_2 removal rate that exceeded their design specifications of 85 percent in 1980 (see table). In addition, each system had a very low energy consumption— 1.6 percent of the total electrical capacity of the boiler. The capital costs of the FGD units were also relatively low, ranging between $42 per Kw and $61 per Kw. Consequently, in spite of the many problems encountered in operating these FGD systems, their overall performance has been successful.

In 1979 the Environmental Protection Agency chose Unit 6 of the Cane Run Station as the demonstration site for a dual-alkali FGD system. EPA is sharing the design, operation, testing and reporting costs of this FGD unit with LG&E. The system was installed at a capital cost of $66 per Kw. The performance data generated from the dual-alkali scrubber have been encouraging; in 1980 it was available for operation 85.9 percent of the time and it had an average SO_2 removal rate of 95.4 percent, despite the high-sulfur (4.8 percent) coal being burned.

LG&E plans to complete another coal-burning power plant in Trimble County by 1985, which will be fitted with an FGD system capable of meeting a strict sulfur dioxide emissions standard of 0.84 lb $SO_2/10^6$ Btu while the plant is using 4 percent sulfur coal.

Waste Disposal

The FGD systems at the Paddy's Run, Cane Run and Mill Creek plants can produce 4,200 tons of FGD waste per day. This waste is treated by the Poz-O-Tec process, which involves dewatering FGD sludge, blending it with fly ash and lime, and depositing the mixture in landfills (see chapter on waste-disposal). The disposal sites for these three plants cover a total of 253.5 acres and will be filled to capacity in seven to ten years. The capital cost of the waste-disposal system was $8 million at Cane Run and $15.7 million at Mill Creek, representing 15 percent of the capital costs for each scrubber.

Table A

LOUISVILLE GAS & ELECTRIC COMPANY
FGD SYSTEMS DATA

	Cane Run 4	Cane Run 5	Cane Run 6
Gross Capacity (Mw)	188	200	299
Percent Utility Ownership	100	100	100
Initial Startup			
Boiler	1962	1966	1969
FGD System	August 1976	December 1977	April 1979
Type of FGD System	Carbide lime	Carbide lime	Carbide lime

Architect/ Engineer	Fluor-Pioneer	Fluor-Pioneer	Fluor-Pioneer
FGD System Supplier	American Air Filter	Combustion Engineering	ADL/Combustion Equipment Assoc.
SO_2 Removal Rate (%)			
Design	85	85	95
Actual (1/80-8/80)	87	90	95.4
Design Particulate Removal Rate (%)	99	99	99.4
Emissions Standards			
SO_2(lb./MBtu)	1.2	1.2	1.2
Particulate (")	0.116	0.116	0.116
Coal Characteristics			
Sulfur (%)	3.75	3.75	4,80
Chloride (%)	0,04	0.04	0.04
Btu/lb.	11,500	11,500	11,000
Energy Consumption			
Percent	1.6	1.5	1.0
Mw	3.0	3.0	3.1
Reagent Requirement(lb/min)	110	125	350 of lime 13.7 of soda ash
Water Requirement (gallons/min.)	80	100	150
Availability (1980)	94.7%	82.7%	85.9%

	Paddy's Run 6	Trimble County 1 & 2
Gross Capacity (Mw)	72	610
Percent Utility Ownership	100	100
Initial Startup		
Boiler	1952	#1 --1985; #2--1988
FGD System	April '73	#1 --July '85; #2 --July '88
Type of FGD System	Carbide Lime	
Architect/ Engineer	Fluor- Pioneer	
FGD System Supplier Supplier	Combustion Engineering	
SO_2 Removal Rate		
Design (%)	90	90
Actual (%)	85	

Design Particulate Removal Rate (%)	99.1	
Emissions Standards		
SO_2 (lb./MBtu)	1.2	0.84
Particulate (")	0.100	RNSPS
Coal Characteristics		
Sulfur (%)	2.50	4.0
Btu/lb.	11,500	
Energy Consumption		
Percent	2.8	1.1
Mw	1.5	7
Reagent Requirement (lb/min.)	75	
Water Requirement (gallons/min.)	20	
Availability (1980)	100%	

	Mill Creek 1	Mill Creek 2	Mill Creek 3	Mill Creek 4
Gross Capacity (Mw)	358	350	442	525
Percent Utility Ownership	100	100	100	100
Initial Startup				
Boiler	1972	1974	1978	1981
FGD System	December '80	December '81	August '78	June '82
Type of FGD System	Carbide Lime*	Carbide Lime*	Carbide Lime	Carbide Lime
Architect/Engineer	Fluor-Pioneer	Fluor-Pioneer	Fluor-Pioneer	Fluor-Pioneer
FGD System Supplier	Combustion Engineering	Combustion Engineering	American Air Filter	American Air Filter
SO_2 Removal Rate				
Design (%)	85	85	85	85
Actual (%)	90-95		86	
Design Particulate Removal Rate (%)	99.4	99.4	99.4	99.4
Emission Standards				
SO_2 (lb./MBtu)	1.2	NSPS	1.2	NSPS
Particulate (")	0.10	0.10	0.10	0.10
Coal Characteristics				
Sulfur (%)	3.75	3.75	3.75	3.75
Chloride (%)	0.06		0.04	
Btu/lb.	11,500		11,500	

Energy Consumption				
Percent	1.4	1.4	1.6	1.4
Mw	5	4.9	7.1	7.2
Reagent Requirement	450 lb/min.		400 lb/min.	
Water Requirement (gallons/min.)	150	150	300	300
Availability (1980)			66.7%	

*Will use lime or limestone after 1984.

Table B

LOUISVILLE GAS & ELECTRIC COMPANY

1979 COSTS

	Cane Run 4	Cane Run 5	Cane Run 6
Capital Cost of Boiler			
Dollars	$20,680,000	$25,000,000	$44,850,000
$/Kw	$110	$125	$150
Capital Cost of FGD System			
Dollars	$12,000,000	$12,000,000	$22,000,000
$/Kw	$60.3	$61.1	$66.0
Annual Costs of Boiler			
Mills/Kwh	1.0	1.0	1.0
Annual Costs of FGD System			
Mills/Kwh	1.188	1.029	7.34
Capital Waste Disposal Costs	$2,500,000	$2,500,000	$2,917,700
Financing	Tax exempt municipal bonds		

	Paddy's Run 6	Trimble County 1 & 2
Capital Costs of Boiler		
Dollars	$7,200,000	
Per Kw	150	
Capital Costs of FGD System		
Dollars	$3,500,000	
Per Kw	$52	

Annual Costs of Boiler Mills/Kwh	1.0
Annual Costs of FGD Systems Mills/Kwh	5.5

	Mill Creek 1	Mill Creek 2	Mill Creek 3	Mill Creek 4
Capital Costs of FGD System				
Dollars	24,000,000	25,000,000	20,000,000	22,000,000
Per Kw	$55.02	$76	$48	$44.4
Annual Costs of FGD System Mills/Kwh			.882	
Capital Waste Disposal Costs	$15,660,000 for all units combined			

NEVADA POWER COMPANY

Fourth Street and Stewart Avenue
P.O. Box 230
Las Vegas, NV 89151

Service Area:	Clark County in southeast Nevada, including Las Vegas.
Regulatory Agencies:	Public Service Commission of Nevada; Federal Energy Regulatory Commission
Electricity Customers (1979):	
Residential	158,692
Commercial and Industrial	18,577
Other	9
Total	177,278
Total (1980)	189,157
Total Electrical Capacity:	1663 Mw
Peak Load:	1403 Mw
Reserve Capacity:	120 Mw from purchases, 18.5%
Sources of Power:	Three coal-burning stations (74%), two natural gas/oil stations (26%), one diesel station, direct purchase from State of Nevada (Hoover Dam).
Coal Source:	Mines in Utah; surface mines in Arizona
Rates (1979):	
Residential	3.17¢ per Kwh
Commercial and Industrial	3.17¢ per Kwh
Other	3.17¢ per Kwh
Total Operating Revenues (1980):	$221,254,627

Source: 1980 Annual Report

The Nevada Power Company (NPC) owns three coal-fired generating plants that provide 74 percent of its electrical capacity. Despite advances in FGD technology over the last ten years, NPC has chosen the same aqueous sodium carbonate scrubber system on three different occasions since 1970. The Reid Gardner station near Las Vegas now has scrubbers on three units, and a fourth is being installed in Unit 4. Depending on future regulations, NPC might be required to retrofit a scrubber on its Navajo plant, which is located near the Grand Canyon. At its Mohave plant emissions comply with state requirements without scrubbers.

The Choice of an FGD System

NPC considered several factors in choosing its scrubber systems for the Reid Gardner station: 1) although NPC burns low-sulfur coals (0.65 percent), it had to meet stringent local emissions standards set in 1973 which limit sulfur dioxide emissions to 0.15 lb/10^6 Btu. (After construction began these standards were relaxed. Currently they must meet a 1.2 lb $SO_2/10^6$ Btu standard.); 2) the architect/engineering firm, Bechtel, examined Reid Gardner Units 1 and 2 and decided that there was room in the plants for scrubbers, but that they had to be located slightly away from the boilers; and 3) all the alternative systems considered and rejected by NPC provided for the recovery of elemental sulfur or sulfuric acid. The possibility of securing an income from the sale of recovered sulfur was discounted by the utility because the coal it burns has such a low sulfur content (0.65 percent) that not enough sulfur would be recovered to justify the expense of a sulfur or sulfuric acid plant.

In 1970 when the scrubbers for Units 1 and 2 were chosen, aqueous sodium carbonate scrubbing was the only process that the utility then considered proven able to achieve a high level (90 percent) of sulfur dioxide removal in a full-sized installation. Aqueous sodium carbonate scrubbing is a wet scrubbing system with an absorbent solution (see Process chapter). Other alternative systems would have required an electrostatic precipitator and an absorber in a configuration that would have been impossible to install given the physical arrangement of the Reid Gardner station, which has no room in front of the air heater (a part of the boiler) for such pollution control equipment.

These and other considerations such as cost, came into play later when the scrubbers for Units 3 and 4 were chosen. Although wet lime and limestone scrubbers use a cheaper reagent, and although these systems had improved by 1974 when the scrubber for Reid Gardner Unit 3 was planned, NPC decided to build another aqueous sodium carbonate scrubber. This was because the savings resulting from the greater reactivity of the sodium carbonate absorbent solution offset the possible cost savings of the cheaper lime or limestone process.

The scrubber on Unit 4, which will also be an aqueous sodium carbonate scrubber, was chosen in 1980. By this time dry scrubbing techniques had been developed but not commercially proven. The utility felt that as dry scrubbers remove at most about 85 percent of the sulfur dioxide, and Unit 4's scrubber must remove at least 85 percent of the sulfur dioxide, a dry scrubber on Unit 4 would, according to the utility, have to operate at "the upper limits of its capability." The cleaned flue gas from a dry scrubber would have picked up as much absorption product as it could possibly hold, which could also cause operating problems.

There were three other considerations in the decision to use this scrubber on Unit 4: 1) since there is a local market for fly ash, and since dry scrubbing could mean contaminated fly ash, NPC preferred to collect the fly ash separately in a baghouse, which would be impossible with a dry system; 2) NPC believed that the cost of the sodium carbonate scrubber was competitive with the cost of a dry scrubber; and 3) the utility was reluctant to try dry scrubbing techniques that had not yet been demonstrated on a full-scale plant.

Construction and Operation

NPC's first two scrubbers were built by Combustion Engineering Associates/Arthur D. Little (now Thyssen-CEA; see profile) and were retrofitted to Reid Gardner Units 1 and 2. Stearns-Roger, Inc. was the architect/engineering firm for the boilers, while Bechtel looked over the vendor specifications and reviewed all of the calculations for the scrubber systems. Bechtel's environmental expertise was the overriding consideration in NPC's decision to hire the firm to design the scrubbers.

However, the involvement of NPC's employees in the design of the scrubbers has increased with the utility's experience. NPC worked directly with the supplier to design the scrubber for Unit 3, while Stearns-Roger engineered the electrical parts and instruments for the scrubber. The utility says that Fluor Power Services was chosen for economic reasons as the architect/engineering firm for the scrubber on Unit 4, which is now under construction. Much of the technical expertise needed to design this scrubber was already available within the utility. The personnel who operated Units 1 and 2 also worked on the design for the scrubber on Unit 3 and made a major contribution to the scrubber design for Unit 4.

Most of NPC's operating experience with the scrubbers has been good. The scrubbers use about 5 percent of the energy produced by the boilers and the availability of three units is high, ranging from 87% to 97% in 1980 (see chart). Because scrubbers are the only means of particulate control, the utility must keep the scrubbers on line to meet the stringent regulations on particulate emissions. However, the scrubbers on Units 1, 2 and 3 are not efficiently removing sulfur dioxide under standards that are presently not strict. Sulfur dioxide removal ranges from 20% (when operating poorly) to 85–90% when operating well.

Units 1, 2 and 3 only have to comply with an emission standard of 1.2 lbs/MBtu and they have complied consistently. However, when

One outside measure of the Reid Gardner station's success is that it has been cited only once for exceeding emissions standards. One reason the utility gives for this success is the fact that all of the coal the unit burns is cleaned. When the coal has been cleaned, less waste is produced by the scrubber and fewer impurities enter the system. Less sludge means a smaller waste disposal problem and fewer impurities mean that the risks of corroding the scrubber parts are lessened.

In addition, when the coal is cleaned, the scrubbed gas can be reheated by mixing it with the gas that bypassed the scrubber and released without violating emission standards. This method of reheating will be used for the scrubber on Reid Gardner Unit 4.

Another reason for the success of the utility's plant is that its machinery undergoes a high level of maintenance. Reid Gardner Units 1-3 require 436 person-hours per month per 100 Mw to run and maintain the scrubbers (see Table B).

Waste Disposal

The waste disposal system at the Reid Gardner station consists of a series of large ponds where the used absorbent water is evaporated. The deposits of calcium salts that are left behind build up until the pond must be drained and the salts carted away. NPC built its ponds with a projected four-year capacity. However, after six years the ponds are not yet full.

Each Reid Gardner scrubber produces 35.4 tons of sludge every day. The Project Manager of Reid Gardner Unit 4 predicted that the ultimate solution for the sludge problem will be recycling the wastes or using them to produce saleable sulfur or sulfuric acid. However, presently there are no plans to recycle any of the FGD wastes from the Reid Gardner station.

The Future of FGD

NPC's response to INFORM's questionnaire identified problems of FGD technology in general, as well as problems specific to the utility's scrubbers, and made some predictions about the future of FGD. The NPC spokesman, David Webster, Generation Engineer, stated that the three problems with FGD technology are generally its overall cost, the disposal of the sludge that it creates, and the fact that FGD systems are sometimes required where they are not what he called "cost-effective." For example, in NPC's case a utility already burning low-sulfur coal is required to spend millions of dollars to remove a relatively small amount of sulfur dioxide from the flue gas.

NPC expects the cost of compliance with emission standards to be high in non-attainment areas where it already has plants. Retrofitting an air pollution control device on the Navajo Plant near the Grand Canyon may be required in order to meet emissions standards for those attainment areas.

Increasing strictness in environmental laws effectively precludes NPC's use of low-sulfur coal as the only means of achieving reductions in emissions, forcing NPC to build scrubbers on all of its future coal-burning plants. For example, when Reid Gardner Unit 4 comes on line in 1983, the current emissions standard of 1.2 lb $SO_2/10^6$ Btu for Units 1, 2 and 3 will be revised for all of Reid Gardners units to a strict standard of 0.15 lb $SO_2/10^6$ Btu.

NPC intends to build scrubbers on the proposed Harry Allen plant near Las Vegas and the proposed Warner Valley plant near St. George, Utah, but the details of these systems have not yet been decided. The construction of these plants has been challenged by the Sierra Club, the Environmental Defense Fund, the Friends of the Earth, who oppose any development involving the mining or burning of coal in southwest Utah or southeast Nevada.

NPC's David Webster predicted that the use of scrubbers will grow in the next 20 years along with the increasing number of coal-burning plants built in the U.S. and abroad. As he sees it, the only alternative to the use of scrubbers is the use of energy sources that don't use coal at all, such as nuclear or solar power. He views the role of the U.S. government as ne of regulator, and says it is the responsibility of the utilities to support research "in order to meet the emissions limits set by the government, and to make electric power plants environmentally safe."

TABLE A

NEVADA POWER COMPANY:
FGD SYSTEM DATA

	Reid Gardner 1	*Reid Gardner 2*
Percent utility ownership	100	100
Capacity	125 Mw	125 Mw
Initial startup		
Boiler	June 1965	June 1968
FGD system	April 1974	April 1974
Type of FGD system	Aqueous sodium carbonate	Aqueous sodium carbonate
Architect/engineer	Bechtel	Bechtel
FGD system supplier	CEA/ADL (now CEA-Thyssen)	CEA/ADL (now CEA-Thyssen)
SO_2 Removal Rate		
Design	90%	90%
Actual	20-90% (est.)	20-90% (est.)
Design particulate removal rate		97%
Emission Standards		
SO_2	1.2 lb/MBtu	1.2 lb/MBtu
EPA		
Particulate	0.1 lb/MBtu	0.1 lb/MBtu
Coal characteristics		
Sulfur	0.65%	0.65%
Chloride	0.05%	0.05%
Btu/lb	11,841	11,841
Energy consumption	5%	5%
Reagent requirement	Reid Gardner Units 1, 2 and 3 use 15 tons/day.	
Water requirement	155 gallons/min	155 gallons/min
Availability (1980)	93.7%	97.5%

	Reid Gardner 3	*Reid Gardner 4*
Percent utility ownership	100	100
Capacity	125 Mw	295 Mw
Initial startup		
Boiler	July 1976	1983
FGD system	July 1976	1983
Type of FGD system	Aqueous sodium carbonate	Aqueous sodium carbonate
Architect/engineer	CEA (now CEA-Thyssen)	Fluor Power Services
FGD system supplier	CEA (now CEA-Thyssen)	CEA (now CEA-Thyssen)
SO_2 Removal Rate		
Design	85%	85%
Actual	68.38 (Oct., 1981)	NA
Design particulate removal rate	99%	--
Emission standards		
SO_2	1.2 lb/MBtu	
EPA	NSPS	
Particulate	0.1 lb/MBtu	
Coal characteristics		
Sulfur	0.65%	
Chloride	0.05%	
Btu/lb	11,841	
Energy consumption	5%	7.3 Mw (estimated)
Reagent requirement	Reid Gardner Units 1, 2 and 3 use 15 tons/day.	
Water requirement	155 gallons/min	
Availability (1980)	87.3%	

TABLE B

NEVADA POWER COMPANY:
1979 COSTS

	Reid Gardner 1	Reid Gardner 2	Reid Gardner 3	Reid Gardner 4
Capital cost of boiler				
In thousands	$15,000	$15,000	$47,000	$289,000
Per kilowatt	$136.36	$136.36	$427.27	$1,156.00
Capital cost of FGD system				
In thousands	$ 5,400	$ 5,400	$14,000	$ 31,000
FPC per kilowatt	$ 48.41	$ 48.41	$126.95	$ 127.20
Annual Costs of FGD				
in 1000's	$1,507	$1,507	$3,906	$8,649 (predicted)
mills/Kwh (FPC)	.8	.94	.99	
Waste disposal costs FPC (mills/kwh)	.02	.02	.02	
Financing	Tax-free pollution control bonds. The scrubbers are exempt from peoperty taxes. No capital expenses are included in the rate base until units are in operation			

PENNSYLVANIA POWER COMPANY
1 East Washington Street
New Castle, PA 16103

Service Area:	Western Pennsylvania
Regulatory Agency:	Pennsylvania Public Utility Commission
Electricity Customers:	
Residential	110,012
Commercial	12,839
Industrial	107
Other	117
Total	123,075
Total Electrical Capacity:	872.2 Mw
Peak Load:	548 Mw
Reserve Capacity:	324.2 Mw
Source of Electric Generation:	97.7% Coal 1.6% Nuclear 0.7% Oil
Rates:	
Residential	5.94¢ per Kwh
Commercial	5.28¢ per Kwh
Industrial	3.37¢ per Kwh
Total Operating Revenues (1980):	$157,280,000

Source: 1980 Annual Report

The Pennsylvania Power Company is the sole operator of the Bruce Mansfield Power Plant, a coal-burning station that is equipped with one of the most successful FGD systems in the United States. In 1980 this FGD system was available for operation well over 90 percent of the time, and it achieved a SO_2 removal rate of 92 percent from the flue gas of boilers using a medium-sulfur (3.07 percent) coal.

Before the construction of the Bruce Mansfield Plant in Shippingport, Pennsylvania, in 1971, a study was conducted by Pennsylvania Power to examine "the state-of-the-art and the expected commercial availability of SO_2 removal systems."[1] Because stringent air pollution regulations (0.6 pounds of SO_2 per million Btu and 0.1 pounds of particulates per million Btu) were soon to be promulgated, the Environmental Protection Agency and the Pennsylvania Department of Environmental Resources would not approve the building of this 2,475

Mw coal plant, consisting of three 825 Mw boilers, without a full-scale scrubbing system. After a worldwide search by Pennsylvania Power for possible FGD suppliers, only Chemico/Envirotech (now General Electric Environmental Services) would accept the contract to equip Bruce Mansfield's first two boilers with lime FGD systems. The reluctance of other suppliers was probably due to their lack of experience in installing scrubbers on such large units.

Installation and Operation

The lime FGD systems on Units 1 and 2 began operating in 1976 and 1977, respectively. Each of these systems consists of two absorbers placed one after the other. The first absorber removes almost all of the fly ash and 70 percent of the SO_2, while the second completes the gas treatment by removing the residual fly ash and another 22 percent of the SO_2.

The FGD system for the third generator, a lime process supplied by M.W. Kellogg Co., began operating in June of 1980. This system removes the fly ash in an electrostatic precipitator first; then the partially cleaned gas passes through a scrubber where 92 percent of the SO_2 is removed. Unfortunately, there is very little operating data on this system.

The operation of the FGD systems on Units 1 and 2 was plagued by several common problems, such as corrosion, plugging and scaling. In order to minimize corrosion from the acidic flue gas, the vulnerable components such as the fans and the smokestack were lined with acid-resistant substances. Plugging and scaling were reduced by using a magnesium-enriched lime and by maintaining acceptable acidity levels. The results of these measures were very good: in the first six months of 1981 both FGD units were available 98 percent of the time the boilers were operating, and both FGD units achieved SO_2 removal rates that exceeded 92 percent.

The energy and labor requirements for the air quality control system at Bruce Mansfield are quite large. The three FGD units use 5 percent of the net electrical capacity of the plant, or a total of 117 Mw. This has served to nearly double the energy consumption of the generating station. In addition, a team of 131 people is required to operate and maintain the complete FGD system, including waste disposal (see table).

The FGD systems at Bruce Mansfield have to remove 92 percent of the SO_2 from the flue gas in order to comply with Pennsylvania's emission standards. These systems have consistently achieved this removal rate for the first six months of 1981. However, Pennsylvania Power has had difficulty meeting the standards, as they require that SO_2 emissions are never to exceed 0.6 pounds of SO_2 per million Btu.

> Variability of equipment operations alone will prevent compliance with this limit on any averaging concept, especially the short term. This does not consider other variabilities the utility has to deal with, such as fuel quality, lime quality, flow rates and load variations.[2]

Using a short-term averaging period of 24 hours instead of a less stringent 30-day averaging period, between March 8 and December 31 of 1979, Unit 1 was in compliance with Pennsylvania's sulfur dioxide emission standards only 58 percent of the time and Unit 2 was in compliance 67 percent of the time.[3]

Waste Disposal

The three FGD units at this plant can produce up to 23,400 tons of slurry (10 to 15 percent solids) per day. Dravo Lime Company designed an elaborate $93 million system to dispose of these wastes. First, the slurry is thickened to a solids content of 30 percent and mixed with Calcilox, a hardening agent (see chapter on waste disposal). The mixture is then pumped seven miles to a 1,400-acre disposal site, created by blocking a steeply walled valley with an earth-and-rockfill dam. The watery sludge will fill the valley, forming a 860-acre lake. In its 15-to-20-year life span, this site will accept approximately 200 million tons of sludge.

Pennsylvania Power has taken several additional steps to minimize the potential environmental impact of the huge quantity of waste water deposited at Bruce Mansfield's disposal site. A portion of the clarified water in the lake is circulated back into the scrubber for reuse. Excess water and runoff are collected in a catchbasin, and water which is clean enough is discharged into the Ohio River. Finally, the disposal area is surrounded by monitoring wells which allow inspectors to determine the quality of the water that leaches from the site.

Table A

PENNSYLVANIA POWER COMPANY
FGD SYSTEM DATA

	Bruce Mansfield 1	Bruce Mansfield 2	Bruce Mansfield 3
% Utility Ownership	4.2	6.8	6.28
Gross Capacity (Mw)	825	825	825
Initial Startup			
Boiler	April 1976	October 1977	October 1980
FGD System	April 1976	October 1977	October 1980
Type of FGD System	Magnesium enriched lime	Magnesium enriched lime	Magnesium enriched lime
Architect/ Engineer	Gilbert Commonwealth Assoc.	Gilbert Commonwealth Assoc.	Gilbert Commonwealth Assoc.
FGD System Supplier	General Electric Environmental Services	General Electric Environmental Services	M.W. Kellogg
SO_2 Removal Rate (%)			
Design	92.1	92.1	92.1
Actual	89	90	95
Design Particulate Removal Rate (%)	99.5	99.5	99.5
Emission Standards			
SO_2	0.6 lb./MBtu	0.6 lb./MBtu	0.6 lb./MBtu
Particulate	0.1 lb./MBtu	0.1 lb./MBtu	0.1 lb./MBtu
EPA	NSPS	NSPS	NSPS
Coal Characteristics (%)			
Sulfur	3.07	3.07	3.07
Btu/lb.	11,373	11,373	11,373
Energy Consumption (%)	5	5	5
Mw	39	39	39
Reagent Requirement	A total of 350,000 tons of lime/year		
FGD Availability	58.2%	76.2%	68%

Costs

According to Pennsylvania Power, "the costs to build and operate [this] FGD lime scrubber are enormously out of proportion with other comparable capital, maintenance and operational costs."[4] The capital cost of the FGD systems, including waste disposal, at the Bruce Mansfield Plant was $409 million, or $175 per Kw; this is equal to 30 percent of the capital cost of the plant. The annual costs of this air quality control system ($30 million) represent 16 percent of the annual costs of the entire generating station.

Table B

PENNSYLVANIA POWER COMPANY

1979 COSTS

Bruce Mansfield Units 1 - 3

Capital Cost of Plant			
Dollars		408,776,000	
Per Kw		175	
Capital Costs of FGD System Without Waste Disposal System			
Dollars		315,476,000	
Per Kw		$135	
Capital Costs of FGD System With Waste Disposal System			
Dollars		408,776,000	
Per Kw		$175	
Annual Costs of Plant			
Mills/Kwh		20.51	
Annual Costs of FGD System			
$/year		30,000,000	
Mills/Kwh	4.72	3.61	3.11
Waste Disposal Costs			
Capital in dollars		93,300,000	
Financing		Pollution control notes; Industrial Revenue Bonds (Beaver County Industrial Development Authority)	

PHILADELPHIA ELECTRIC COMPANY

2301 Market Street
P.O. Box 8699
Philadelphia, PA 19101

Service Area:	2,475 square miles in southeastern Pennsylvania (including Philadelphia) and northeastern Maryland; population 3,700,000.
Regulatory Agencies:	Pennsylvania Public Utility Commission; Federal Energy Regulatory Commission; Nuclear Regulatory Commission; and other state, federal and local agencies including the Pennsylvania Department of Environmental Resources.
Electricity Customers:	
Residential	1,190,312
Small Commerical and Industrial	116,808
Large Commercial and Industrial	5,820
Other	736
Total	1,313,676
Total Electrical Capacity:	7,698 Mw
Peak Load:	6,095 Mw
Reserve Capacity:	Provided by purchases and interchange, 26.3%
Sources of Electrical Capacity (1980 estimates):	
Nuclear	30.5%
Service Area Coal	12.7
Mine-Mouth Coal	11.5
Oil	13.3
Hydro	4.3
Internal Combustion	1.6
Purchase and Net Interchange	26.0
Coal Source:	Long-term contracts for bulk of requirements; spot market purchases.
Rates:	
Residential	5.98¢ per Kwh
Small Commercial and Industrial	6.45¢ per Kwh
Large Commercial and Industrial	3.81¢ per Kwh
Other	4.77¢ per Kwh
Total Operating Revenues(1980):	$2,123,394,000

Source: 1980 Annual Report

The Philadelphia Electric Company (PE) acquires about one-fourth of its electrical capacity from coal-fired generating stations. Two of these, the Cromby and Eddystone plants, which account for slightly less than half of the utility's coal capacity, have planned FGD systems that will be the only magnesium oxide scrubbers in the U.S. The utility burns 2.55 percent medium-sulfur coal, and must meet an emissions limit of 0.9 lb SO_2/MBtu at the Cromby Station and 0.45 lb SO_2/MBtu at the Eddystone station.

The Choice of an FGD System

After new regulations were implemented by the Pennsylvania Department of Environmental Resources in the early 1970s, Philadelphia Electric was required to retrofit sulfur dioxide and particulate scrubbers to its Eddystone and Cromby coal-burning power plants. The utility chose an experimental technology that uses magnesium oxide as a regenerable absorbent (see discussion in Process chapter) because the scrubbers are required to remove more than 92 percent of the sulfur dioxide and because there is limited space available for waste disposal.

Construction and Operation

Working with a senior engineer of the utility, United Engineers and Constructors designed and built a one-third sized prototype scrubber for Philadelphia Electric's Eddystone Unit 1 power plant. The prototype scrubber was tested and refined between 1976 and 1978, and in 1979. A full-sized magnesium oxide scrubber is being installed in place of the prototype and is scheduled to go into operation in December 1982. Construction is also under way on identical FGD systems on the Cromby Unit 1 power plant and on the second Eddystone unit. PE is using cleaning coal to meet interim emissions standards until the scrubbers are completed, when it will revert to medium-sulfur coal (2.55 percent).

Performance data will not be available until the scrubbers begin commercial operation. However, after testing the prototype scrubber, the utility concluded that the commercial units will be capable of meeting Pennsylvania's emissions standard of 0.45 lb SO_2/MBtu. In fact, the major problems encountered by the scrubber's operators during testing were mechanical. In a joint report to the EPA symposium on

FGD in November 1977, the utility and the architect/engineering firm concluded that "no fundamental problems involving the scrubber chemistry have been observed."

As construction is proceeding on the units, it seems safe to infer that the utility is satisfied with the system. The same 1977 report also points to the prospect of some savings derived from the experience gained during the prototype testing. As a result of this testing, equipment can be simplified, and less equipment will be needed than previously thought. However, INFORM was unable to obtain financial data and no exact figures are available.

TABLE A

PHILADELPHIA ELECTRIC COMPANY
FGD SYSTEM DATA

	Cromby	Eddystone 1	Eddystone 2
Percent utility ownership	100	100	100
Capacity	150 Mw	310 Mw	334 Mw
Initial startup			
Boiler	1954	1959	1960
FGD system	May 1983	December 1982	December 1982
Type of FGD system	Magnesium oxide	Magnesium oxide	Magnesium oxide
Architect/engineer	United Engineers and Constructors	United Engineers and Constructors	United Engineers and Constructors
FGD system supplier	United Engineers and Constructors	United Engineers and Constructors	United Engineers and Constructors
Design sulfur dioxide removal rate	95%	92%	92%
Design particulate removal rate	90%	90%	90%
Emission standards			
Sulfur dioxide	0.9 lb/MBtu	0.45 lb/MBtu	0.45 lb/MBtu
Particulate	0.1 lb/MBtu	0.1 lb/MBtu	0.1 lb/MBtu

Coal characteristics			
Sulfur	2.55%		
Chloride	0.14%		
Btu/lb	12,673		
Energy consumption*	2.0%	1.93%	1.80%
Reagent requirement	18 to 20 tons of magnesium oxide per day for each 100 Mw. MgO costs $250 to $300 per ton.		
Water requirement	450 gpm	1,000 gpm	1,000 gpm
Availability	----	----	----

*without regeneration facility

Costs

Two site-specific factors will have an effect on the cost of PE's FGD system. (Here again, no exact figures are available.) First, PE's solid scrubber product, magnesium sulfite, will be dried and trucked to a Newark plant to be used in the production of sulfuric acid. Magnesium oxide will also be regenerated at the plant. The utility states that its energy use and capital costs will be lowered by using this already existing sulfuric acid plant.

Second, because the Eddystone station is located close to the Philadelphia airport, the plant cannot emit a visible plume of smoke. Consequently, the flue gas temperature must be raised by 100° F (rather than the normal 20° F to 50° F) so that all of the water droplets are vaporized. This will increase operating costs by about 10 percent.

The Future of FGD

Based on their experience with an experimental scrubber, representatives of PE cited instrument problems, the harsh and corrosive environment inside a scrubber, and scrubber availability as the major problems of FGD technology. These representatives see FGD only as an interim technology that will be replaced in the next five to ten years by fluidized-bed combustion, solvent refined coal and liquefaction, and gasification.

The representatives of PE interviewed by INFORM said that the government should try to create a more positive regulatory environment, where investment in pollution control technologies is encouraged. They also expressed an interest in exploring a regulatory structure that gives some kind of credit to a utility that removes more sulfur dioxide than emissions standards require, and suggested that the government might not play as important a role in the future as it has in the past in the development of FGD.

TABLE B

PHILADELPHIA ELECTRIC COMPANY

1979 Costs

	Cromby	Eddystone 1	Eddystone 2
Capital cost of boiler			
In thousands	NA	NA	NA
Per kilowatt	NA	NA	NA
Capital cost of FGD system			
In thousands			
$/kW (FPC)	$533/Kw	combined $357/Kw	
Annual revenue requirements of boiler	NA	NA	NA
Estimated Levelized annual revenue requirements of FGD system	36.5 mills/Kw	23.1 mills/Kw for Eddystone 1 and 2	
Waste disposal costs	$67/Kw for each unit (regenerable system)*		
Financing	NA	NA	NA

NA = Not Available

*Source: FPC form #67.

PUBLIC SERVICE COMPANY OF NEW MEXICO
Alvarado Square
Albuquerque, NM 87158

Service Area:	North-central New Mexico, including Albuquerque; population 565,000.
Regulatory Agency:	Public Service Commission of New Mexico
Electricity Customers:	
Residential	191,495
Commercial and Industrial	21,398
Other (including resale)	264
Total	213,157
Total Electricity Capacity:	1,080 Mw
Peak Load:	913 Mw
Reserve Capacity:	Purchases and peaking stations, 18.3%
Sources of Power:	Two coal-fired stations, 756 Mw (72%); Three gas- and oil-fired stations, 291 Mw (28%)
Coal Source:	Local coal mined by Utah International
Rates (1979):	
Residential	5.928¢ per Kwh
Commercial	5.258¢ per Kwh
Industrial	4.466¢ per Kwh
Other	5.197¢ per Kwh
Total Operating Revenues (1980):	$280,516,000

Source: 1980 Annual Report

The Public Service Company of New Mexico (PNM) derives 72 percent of its electricity from two coal-fired power plants. One of these plants, the San Juan station, burns low-sulfur coal (0.8 percent) and has a large Wellman-Lord scrubber which is designed to remove 90 percent of the sulfur emissions. PNM shares the ownership of the San Juan plant Units 1, 2, and 3 with the Tucson Electric Power Company. PNM operates the plant. PNM has a 13 percent interest in the second station, the Four Corners plant, operated by Arizona Public Service Company.

The Choice of an FGD System

PNM's interest in flue gas desulfurization is a result of state legislation which currently mandates strict sulfur dioxide emissions limits of 0.65 lb/10^6 Btu, 60 percent SO_2 removal, and 1.2 lb SO_2/10^6 Btu for its San Juan Units 1, 2, 3, respectively. The utility, in response to the INFORM questionnaire, stated that "the citizens of New Mexico indicated that they were willing to pay the price for clean air, therefore, PNM attempted to provide the cleanest available power."

Thus when the utility chose a scrubber in 1974 it had to find one that could achieve a very high rate of sulfur dioxide removal. The utility chose a Wellman-Lord system (see description in Process chapter) that was not yet commercially proven. PNM was not the first utility to build a Wellman-Lord scrubber, since a full-sized Wellman-Lord scrubbing system had been in use at Northern Indiana Public Service Company's Dean H. Mitchell 11 plant since 1976. However, PNM's San Juan Units 1 and 2 were the next to come into operation, so the utility's choice was somewhat experimental.

PNM chose its Wellman-Lord scrubber for several reasons. First, as mentioned above, it is designed to achieve a high rate of sulfur dioxide removal of 90 percent. Second, PNM operates in an arid area in the southwest, and its Wellman-Lord system uses less water (365 gallons per minute) than a lime or limestone system of comparable size (which can use about 560 gallons per minute). Third, as the San Juan system has no railroad, the utility is limited to the expensive option of trucking. PNM needed a system requiring the least transportation of materials, and the Wellman-Lord's regenerable reagents avoids the problems of continuous shipments into the plant. Fourth, although PNM also considered such proven scrubbers as the limestone and dual-alkali system (see Process chapter), it chose the Wellman-Lord because it felt that the system provided comparable qualities, what it called "redundancy and sophistication," at the lowest possible cost.

Construction and Operation

Before it gained operating experience and gathered operating and chemical teams of its own, PNM was very dependent on its A/E firm, Stearns-Roger, to run its scrubbers when they started operating in 1978. Like the FGD systems of other utilities, PNM's Wellman-Lord scrubber has had its share of start-up and operating problems. These difficulties included: the loss of absorbent solution, the malfunction of the electrostatic precipitators, problems with the circulation of system solutions, the improper installation of the equipment in the Claus plant sulfur-generating system, and insufficient supplies of water and steam.

Some of the solutions to these problems have involved increasing the number of operators, redesigning or re-installing equipment in the system, and developing improved operating procedures. PNM now has a staff of about 100 members who operate and maintain the scrubber and who perform administrative tasks.

TABLE A

PUBLIC SERVICE COMPANY OF NEW MEXICO
FGD SYSTEM DATA

	San Juan 1	*San Juan 2*
Percent utility ownership	50	50
Capacity (Mw)	361	350
Initial startup		
Boiler	December 1976	November 1973
FGD system	April 1978	August 1978
Type of FGD system	Wellman-Lord	Wellman-Lord
Architect/engineer	Stearns-Roger	Stearns-Roger
FGD system supplier	Davy Powergas	Davy Powergas
SO_2 Removal Rate		
Design	90	90
Actual	60	60
Design particulate removal rate	99.8%	NA
Emission Standards		
SO_2	0.65 lb/MBtu	60% (reduction rate)
EPA	NSPS	NSPS
Coal Characteristics		
Sulfur	0.8%	0.8%
Chloride	0.03%	0.03%
Btu/lb	9,800	9,800

Energy consumption (Mw)	25	25
Reagent requirement	NA	NA
Water requirement	365 gallons/min	355 gallons/min
Availability (1980)	91.6%	83.5%

	San Juan 3	San Juan 4
Percent Utility Ownership	50	50
Capacity (Mw)	534	472 (estimated)
Initial Startup		
Boiler	December 1979	1982
FGD System	December 1979	1982
Type of FGD System	Wellman-Lord	Wellman-Lord
Architect/Engineer	Brown & Root	Brown & Root
SO_2 Removal Rate		
Design	90%	90%
Actual	30	
Design Particulate Removal Rate	99.5%	NA
Emission Standards		
SO_2	1.2 lb./MBtu	1.2 lb./MBtu
EPA	NSPS	RNSPS
Coal Characteristics		
Sulfur (%)	0,8	0.8
Chloride (%)	0.03	0.03
Btu/lb.	9,800	9,800
Energy Consumption (Mw)	25	25
Availability (1980)	83.8%	NA

NA: Not available

Costs

PNM sources state that 12.7 percent of a customer's bill pays for pollution controls, the bulk of which are FGD-related. The utility estimates that the cost of the scrubber system has exceeded estimates by $60 million (this figure is not adjusted for inflation). PNM stated that the capital cost of the scrubbers on Units 1 and 2 was $120 million, but did not state whether the $60 million is included in this figure. It costs the utility $6 million each year to run the scrubber system.

Table B

PUBLIC SERVICE OF NEW MEXICO

1979 Costs

	San Juan 1	San Juan 2	San Juan 3	San Juan 4
Capital Costs of Boiler				
in 1,000's	$101,770	$79,509	$277,390	$275,072
Per Kw	$577	$548	$578	$583
Capital Cost of FGD System				
in 1000's	$120,000 for Units 1 and 2		$96,946	$90,039
Per Kw	$132.81	$137.1	$185.03	$224.32
Annual Revenue Requirements of Boiler	7,500 tons of coal per day at $15 per ton			
Annual Costs of FGD $/year	$6,000,000 per year for Units 1 and 2		NA	NA
Waste Disposal Costs	Regenerable system allows for the sale of sulfur. Expect $2,000,000 per year, as well as $60,000 per year in state sales tax.			
Financing	All funds for scrubbers received through pollution control revenue bonds issued by the city of Farmington, New Mexico.			

NA: Not available.

The Future of FGD

The utility directed much of the blame at the government for the most significant problems of flue gas desulfurization, including severe operating problems that force the scrubber to shut down, difficulties in the disposal of FGD wastes, and high capital and operating costs. "The Clean Air Act forced untried technologies into use before they were commercially proven with many not unexpected operational problems as a result," PNM stated in its response to the INFORM questionnaire.

PNM stated that the proper role of government should be "in research and development programs that advance FGD technology and that solve some of the present-day problems with existing technology." PNM believes that in the future, regulators should not force utilities to invest large sums of money in technologies whose long-term performances have not been demonstrated; rather utilities should support such institutions as EPRI, as they work to develop control technologies.

However, PNM does believe that FGD systems will be a necessary part of this country's energy plans as long as both the use of coal and the public awareness and concern about its impact on the environment increase. In the long run, beyond the next ten years, PNM believes that the technology will become important internationally as well.

The utility is vague about the question raised by the technology-forcing aspect of the Clean Air Act. If the government had not set standards beyond the reach of commercially available technology, the technology probably would not have been developed. PNM cited the Japanese Chiyoda Thoroughbred and DOWA processes (see discussion in Process chapter) as examples of promising processes that were developed at no cost to American industry. Like other utilities however, PNM would be reluctant in the future to purchase systems such as those mentioned above that had not been commercially proven.

Appendix

Appendix

The Clean Air Act and FGD

The broad purpose of the Clean Air Act (CAA) is "... to protect and enhance the quality of the nation's air resources so as to promote the public health and welfare. . . ."* Although the Act was passed in 1955, it was not until 1970 that the federal government gave the power to the Environmental Protection Agency (EPA) to set uniform national air quality standards, and some states and localities set pollution control standards on their own, which in some cases were stricter than the federal standards.

New Source Performance Standards

Although the regulations promulgated in 1970 by EPA under the CAA cover many pollutants emitted by most major industrial sources, presently scrubbers are neede primarily to control sulfur dioxide emissions from major utility plants. The 1970 New Source Performance Standards (NSPS) required that all major new and modified industrial

---

*Clean Air Act—Section 101 (b) (1)

sources meet minimum emission standards for specific pollutants set by EPA. EPA set these standards beyond the reach of pollution control equipment then commercially available specifically to force the development of flue gas desulfurization technology. Utility coal-fired boilers of 73 Megawatt output or greater, on which construction or modification had begun after August 17, 1971, could not emit more than 1.2 pounds of sulfur dioxide (SO_2) per million (10^6) Btu's (British thermal units) of boiler heat output. Plant operators were required to use "continuous emission monitoring" (CEM) to measure the sulfur dioxide emission levels in the flue gas outlet. If the average emissions exceeded the NSPS standard for more than three hours, the plant was liable to be cited in violation.

The EPA arrived at this nationwide standard of 1.2 lb $SO_2/10^6$ Btu by assuming:

1. the use of coal with an average sulfur content of 4 percent (generally the percentage found in eastern and midwestern coals), and
2. that the best available scrubber pollution control technology could remove 70 percent of the emissions from a coal with a heating power of one million Btu per pound.

By 1977, however, the provisions of the CAA were being strongly contested. Utility operators complained that the standards were too strict and that FGD systems were still untested and unable to meet them (only three scrubbers had been in operation by 1971). The eastern and midwestern coal companies argued that the Act threatened their markets by encouraging utilities to buy low-sulfur coal from western mines. Environmentalists argued that the standards were not strict enough, allowing coal-burning plants to go only as far as meeting the standard by using low-sulfur coal.

The amendments to the Clean Air Act in 1977 were achieved after a curious political merger: the midwestern and eastern coal operators joined the environmentalists to insist on requiring *all* coal burning plants to adopt the only technology capable of reducing sulfur dioxide emissions by at least 70 percent. Thus the western coal operators would be prevented from selling large quantities of low-sulfur coal in the eastern market, and utilities would have to adopt scrubbers for new power plants even if they could meet the NSPS standards by using low-sulfur coal.

Revised New Source Performance Standards

In 1979, after receiving oral and written comment from business, conservation and consumer groups, EPA promulgated its Revised New Source Performance Standards (RNSPS). These were stricter than the NSPS and applied to all coal burning plants capable of producing more than 250 BTU an hour, (or about 73 Megawatts of generating capacity) on which construction or modification had begun after September 12, 1978.

The new regulations retain the 1971 NSPS standard of 1.2 lb SO_2/ 10^6 Btu as a ceiling for emissions, but additionally require that sulfur dioxide emissions from all new (post-1978) boilers be reduced on a percentage sliding scale that takes into account the differences in sulfur content of U.S. coals. All coals burned must have at least 90 percent of the sulfur dioxide removed from their smoke, unless 90 percent removal reduces emissions to less than 0.6 lb/10^6 MBtu. If emissions go below this "floor" of 0.6 lb., reductions between 70 and 90 percent are permitted, depending upon the sulfur content of the coal. In addition, utilities are required to continuously monitor sulfur dioxide emissions both at the flue gas inlet and at the outlet of these new sources to determine whether the scrubber attains the RNSPS percentage removal on a 24-hour rolling average.

The effect of these RNSPS regulations is that flue gas desulfurization is now required on all new utility boilers, since only scrubbers can achieve over 70 percent reduction in emission.

All these standards require that a scrubber must continually correct for changing amounts of sulfur dioxide in the boiler stack as emissions vary. Fluctuations in the sulfur content of the coal can vary by a percentage point or more, changing the amount of sulfur dioxide that is emitted and which must be removed. Fluctuations in the daily electrical demand of the boiler also affect the amount of flue gas that is flowing through the FGD system and hence the total amount of sulfur dioxide produced. High emissions during one part of a time period can be used to offset low emissions during another part of the period over which total emissions are averaged.

National Air Quality Standards

Another section of the CAA may also vary the requirements for pollution control that boilers have to meet. In 1970, the CAA had mandated that EPA set National Ambient Air Quality Standards (NAAQS) for seven of the most common and widespread pollutants, of which sulfur dioxide is one. The entire country was divided into 274 air quality control regions (AQCRS), and the NAAQS had to be met in all of them. Control regions within state boundaries that did not meet all of the NAAQS were called "nonattainment areas" and states are required to devise a strategy to insure that the minimum standards set by EPA are met and maintained. Under these State Implementation Plans, or SIPs, new and modified sources within nonattainment areas are required to achieve the "lowest achievable emission reduction" (LAER), regardless of cost.

In 1977 the CAA amendments also required the states to set limits on the *existing* major pollution sources within nonattainment areas. It was specified that such sources must use "reasonably available pollution control technologies" (RACT). Both technological and economic feasibility are considered when applying RACT to existing sources.

In "attainment areas," or control regions where NAAQS *have* been met, new and modified pollution sources were regulated to "prevent significant deterioration" (PSD) of the clean air within the control regions. These sources were required to use the "best available control technology" (BACT). BACT is an emission limitation based on the maximum degree of reduction that can be achieved when energy, environmental, and other costs are considered.

In other words, if air quality is good, the CAA says it should remain good; clean air should be protected. Some increase in air pollution is permitted, but not as much as would be allowed if NAAQ standards were applied. In national parks or wilderness areas, the allowable increase in pollution is even less than in other areas, and no impairment of visibility from pollution is allowed. In such regions, state regulations, especially for new development, may be very strict.

Even in air quality regions with the same federal ambient air requirement, state and local air pollution standards can vary. Not only can local standards be more strict than the federal standards, but different state implementation plans allow different methods for meeting the federal air quality standards. Different localities must respond to variable local conditions, such as different levels of industrial development and traffic, different existing levels of air pollution, or different weather conditions. Thus boilers and other emission sources close to each other may have very different emission standards to meet.

Also, utilities with both old and new boilers meet different standards. For example, Duquesne Light has boilers near Pittsburgh in a nonattainment area. Although these are old (pre-1978) boilers, not subject to RNSPS, they must meet a state emissions limit for sulfur dioxide of $0.6/10^6$ Btu under the SIP. Similarly, at Associated Electric's Thomas Hill plant in Moberly, Missouri, which is in an attainment area, station Units 1 & 2 have old boilers that have only to meet the lenient state standard of sulfur dioxide removal of 9.5 $lbs/10^6$ Btu. But Unit 3, at the same location, is a new boiler and must meet the federal NSPS of 1.2 $lbs/10^6$ Btu.

Two of the foregoing requirements—LAER and BACT—would require the use of scrubbers on new and modified sources of pollution in both nonattainment and attainment areas. But the impact of these regulations will not be felt until the states complete their state implementation plans. EPA set a 1975 deadline for such plans and another deadline for 1979, but by mid-1980 only two SIPS had been fully approved, and EPA has extended the deadlines so that in some cases the SIPs won't have to be completed until 1987.

By 1981, when Congress began reviewing the CAA, utilities and other business interests were pressing to have the CAA further amended so as to eliminate LAER, PSD and BACT, as well as the minimum standards EPA has set for the SIPS.

Notes

Introduction Notes

1. U.S. Environmental Protection Agency, *1977 National Emissions Report* (March 1980).

2. Gene E. Likens, Richard F. Wright, James N. Galloway, Thomas J. Butler, "Acid Rain," *Scientific American,* Vol. 241, No. 4, October 1979, pp. 43-51.

3. Edison and Walker, "A Review of Sulfur Dioxide and Particulate Matter as Air Pollutants with Particular Reference to Effect on Health in the United Kingdom," *Environmental Research,* Vol. 16, 1978, p. 302; Fishelson and Graves, "Air Pollution and Morbidity: SO_2 Damages," *Journal of the Air Pollution Control Association,* Vol. 28, 1978, p. 785.

4. National Academy of Sciences, *Atmosphere-Biosphere Interactions: Toward A Better Understanding of the Ecological Consequences of Fossil Fuel Combustion* (Washington, DC: National Academy Press, October 1981).

5. Reported in *Ecology Law Journal,* Vol. 9, No. 1, p. 89.

6. As reported in *Mining Congress Journal,* Vol. 67, No. 9, September 1981, p. 65.

7. PEDCo Environmental Inc., "Categorical Summaries of FGD Systems," *EPA Utility FGD Survey: October-December 1980, Vol. 1* (EPA-600/7-81-021a, January 1981).

8. Ibid., p. xx.

9. William T. Lorenz & Co., *1981 Update-Air Pollution Control Industry Outlook,* and verbal estimates.

Findings
Notes

1. Availabilities were calculated from data obtained from PEDCo Environmental Inc., *EPA Utility Survey: October-December 1980, Vol. 2* (EPA-600/7-81-012, January 1981), pp. 47-656.
2. Estimate by Komanoff Energy Associates, New York City, 1981.
3. Bechtel National, Inc., *Economic and Design Factors of the Flue Gas Desulfurization Technology* (Prepared for Electric Power Research Institute: CS-1428, April 1980).
4. *New York Times*, May 5, 1981, p. D1.

FGD Systems and How They Work
Notes

1. PEDCO Environmental, Inc., "Categorical Summaries of FGD Systems," *EPA Utility FGD Survey: October-December 1980, Vol. 1* (EPA-600/7-81-012a, January 1981), p. 661.
2. PEDCo Environmental, Inc., "Design and Performance Data for Operational FGD Systems," *EPA Utility FGD Survey: October-December 1980, Vol. 2* (EPA-600/7-81-012b, January 1981), pp. 4-19.
3. Personal communication with Randall Rush, Southern Company Services, August 6, 1981.
4. *Pollution Engineering*, Vol. 12, No. 11, November 1980, p. 4.
5. PEDCo Environmental Inc., *EPA Utility FGD Survey Vol. 1*, p. A-19.

The Treatment and Disposal of Waste from Wet Lime and Limestone FGD Systems
Notes

1. Chakra J. Santhanam and Julian W. Jones "Characterization and Environmental Monitoring of Full-Scale Utility Waste Disposal: A Status Report" *Proceedings: Symposium on Flue Gas Desulfurization—Houston—October, 1980, Vol. 2* (U.S. Environmental Protection Agency: EPA-600/9-81-019b), p. 572.
2. Environmental Protection Agency, Industrial Environmental Research Laboratory, *Capsule Report: Disposal of Flue Gas Desulfurization Wastes—Shawnee Field Evaluation* (U.S. Environmental Protection Agency: EPA-625/2-80-028, October, 1980).
3. Michael Baker Jr. Inc., *FGD Sludge Disposal Manual* (Palo Alto, CA: Electric Power Research Institute, January, 1979).

4. Ibid.

5. Environmental Research and Technology, Inc., *Coal Mine Disposal of Flue Gas Cleaning Wastes* (Palo Alto, CA: Electric Power Research Institute, June 1980).

6. P.M.J. Woodhead, J.H. Parker and J.W. Duedall, "Environmental Compatibility and Engineering Feasibility for Utilization of FGD Waste in Artificial Fishing Reef Construction," *Symposium Flue Gas Desulfurization,* pp. 695-700.

7. *Environment,* Vol. 23, No. 1, Jan./Feb. 1981, p. 24.

8. J.D. Veitch, A.E. Steel and T.W. Tarkington, *Economics of Disposal of Lime/ Limestone Scrubbing Wastes: Surface Mine Disposal and Dravo Landfill Processes* (Prepared for U.S. Environmental Protection Agency by Tenessee Valley Authority—EPA-600/7-80-022).

Louisville Gas & Electric Company Notes

1. *Synchronizer* (L.G.&E., pub.), Vol. 58, No. 3, Fall 1979.

Pennsylvania Power Company Notes

1. R.C. Forsythe, "Operating Experiences—Bruce Mansfield Plant Flue Gas Desulfurization System," Paper presented at Pennsylvania Electric Association Systems Operation Committee Meeting, Sharon, Pennsylvania, October 23, 1979.

2. C.V. Runyon and R.C. Forsythe, "An Update on Flue Gas Scrubber Experiences—Bruce Mansfield Plant," Paper presented at Utility SO_2 Scrubber Conference, March 25–28, 1980, p.16.

3. Ibid., p. 18.

4. Ibid., p. 31.

Methodology

Literature Review

This project was initiated in November 1979, with a ten month review of literature on scrubbers gathered from four primary areas:

Government documents from the U.S. EPA and DOE and State Public Service Commissions.

Industry materials available from individual corporations producing scrubbers, from utilities using them, from relevant trade associations and from research centers including the Electric Power Research Institute, the Edison Electric Institute, the National Coal Association and the Mine and Reclamation Congress.

Business and general publications ranging from industry periodicals such as *COAL AGE, COAL WEEK,* to the *Wall Street Journal* and *New York Times.*

Publications from independent research centers.

Selection of Project Advisory Board

A project advisory board was established, including representatives from business, government, public interest and independent research sectors. The role of this Board was to offer advice on the directions of the project, on the extent of research to be carried on, and on the final report manuscript. Extensive consideration was given to comments and advice of Board members (see list of advisors). However, the final responsibility for the content of this study resides with the authors and INFORM.

Questionnaires

Questionnaires for the utilities and the scrubber supplier firms that INFORM planned to include in this study were developed, defining the information INFORM would be seeking on FGD systems, their problems and successes and costs. The questionnaires were reviewed by advisors before being finalized.

Selection of Sample Firms

11 of the 16 FGD suppliers in the U.S. selling the most systems were selected as potential participants for this study based on their experience in the field, including experience with some of the newer FGD systems. The selection was made to include suppliers of every current scrubber system in use.

When these suppliers were asked to participate in the study, ten agreed to do so. The eleventh, Davy McKee Co., was reluctant to participate but agreed to be interviewed by phone. The ten companies were sent INFORM's questionnaire for suppliers, and personal interviews were conducted with R&D and marketing vice presidents at each firm during September, October and November 1980.

14 utilities currently using or planning to use scrubbers were also selected for participation in this study. INFORM's selection sought to include utilities with a range of experience in using various scrubber types including some who have experience with the newer type of systems:

- seven have been operating conventional lime/limestone FGD systems for up to nine years;
- five operate several kinds of FGD systems on a variety of their boilers; and
- seven operate systems other than the conventional lime and/limestone types of scrubbers.

Research Procedure

INFORM contacted and sent questionnaires to the 14 utilities and ten of these agreed to participate in the study. Of these ten, INFORM obtainted sufficient information from seven utilities with which to write a profile of each.

The four utilities that did not contribute information are: Texas Utilities, Northern State Power Company, Central Illinois Public Service Company, and Northern Indiana Public Service Company which was undergoing a strike when cooperation was requested. The three utilities that gave inadequate information were: TVA, Arizona Public Service Company and Kansas Power and Light Company.

In-depth interviews were conducted with utility executives and managers of environmental affairs, pollution control and scrubber operations in nine of the ten companies that participated. One company, Pennsylvania Power, sent INFORM technical papers from which its profile was written.

Report Preparation

Profiles were written, reviewed in-house and then sent back to the respective companies to check for accuracy. Upon receiving the company comments, the profiles were revised and then edited by INFORM's staff.

The information for the background chapters on FGD processes and waste disposal was obtained, for the most part, from the literature.* These chapters were reviewed by the project advisors and revised by the authors.

Writing and Editing

The findings were written from information obtained from the interviews conducted by INFORM and from the current literature. Once again, these findings were reviewed by our project advisors and revised by INFORM's staff. INFORM's staff accepts full responsibility for the final conclusions drawn from the research.

---

*Note: Many references are footnoted.

Project Director and Author

Mary Ann Baviello joined INFORM's energy staff in January 1980. She has been a science intern at the National Resource Defense Council, researching the biological effects of microwave and radio-frequency radiation, and a research assistant in the Life Sciences at the University of Nebraska. She received a B.A. with Honors in Environmental Science and served as a teaching and administrative assistant at the State University of New York, College at Purchase. She is currently pursuing graduate work in the field of environmental management and protection at the University of North Carolina at Chapel Hill.

Research Assistants

Alexandra S. Bowie joined the INFORM energy staff in January 1981 after working on the Issues Staff of Mark Green for Congress, researching energy policy and other political issues. Ms. Bowie also researched energy policy and wrote position papers for the Kennedy for President campaign. She received her B.A. from Bryn Mawr College, has been a law assistant, and is currently a candidate for a law degree at Boston University.

Lillian E. Beerman joined INFORM's energy staff in November 1981 after receiving a Master's of Forest Science from Yale's School of Forestry and Environmental Studies. Her work at Yale concentrated on energy and environmental regulation and resource planning. She is a co-author of "Solar Technologies," which is included in *Assessing the Social Dimensions of Renewable Energy Systems,* a working paper published by the Yale School of Forestry and Environmental Studies. She has worked as a graduate intern at the Solar Energy Research Institute, researching and writing cost benefit analyses. She previously was a congressional intern for Representative Robert Eckhardt, and is currently on the research staff of Senator Lloyd Bentsen.

INFORM PUBLICATIONS LIST

Newsletter

INFORM Reports: A bi-monthly newsletter reporting on INFORM's current research and educational activities. Articles and announcements.
$25 per annum (membership fee)

Air and Water Pollution

A CLEAR VIEW: Guide to Industrial Pollution Control (June 1975)
A manual on procedures for monitoring and assessing industrial air and water pollution problems and controls.
$6.95/$3.95 paperback

Nutrition

WHAT'S FOR DINNER TOMORROW? Corporate Views of New Food Products (January 1981)
An analysis of where, when, and to what degree nutritional considerations are factored into corporate decisions to develop and market a new food product.
$5.00

Energy

RECLAIMING THE WEST: The Coal Industry and Surface-Mined Lands (July 1980)
An analysis of the land reclamation practices of 13 coal companies at 15 surface mines in 6 western states. Defines best available reclamation techniques and evaluates the impact of each mine upon local land and water resources.
$45.00

ENERGY FUTURES: Industry and the New Technologies (July 1976)
An analysis of corporate research, commercial development, and environmental impact of new energy sources and technologies. (check availability before order)

INDUSTRIAL ENERGY CONSERVATION: Where Do We Go From Here? (December 1977)
A report on the federal programs and industry progress toward achieving federal energy conservation targets.
$5.00

FLUIDIZED-BED ENERGY TECHNOLOGY: Coming To A Boil (June 1978)
A report on the state-of-the-art of fluidized-bed combustion technology for cleaner and more efficient direct burning of coal. This report provides information on the companies now researching and developing fluidized-bed systems, the variety of possible large- and small- scale applications for the technology, its cost, and the barriers to widespread commercial use.
$45.00

CLEANING UP COAL: A Study of the Technology and Use of Cleaned Coal (March 1982)
A close examination of various coal-cleaning techniques, reporting on the state-of-operation and state-of-the-art of coal cleaning systems. The study examines the present status of sulfur- and ash-removing capabilities of preparation plants, costs and environmental effects of clean coal processes and utilities' experiences with burning cleaned coal. The study profiles experiences of 28 companies in four different areas connected with the technology: construction, ownership, operation, and use of this fuel. Reports also on the status of research and development of coal cleaning systems by the DOE, EPA and EPRI.

THE SCRUBBER STRATEGY: The How and Why of Flue Gas Desulfurization (April 1982)
An examination of the state-of-operation and state-of-the-art of flue gas desulfurization technology, including the various systems available; the problems of waste disposal and treatment methods; the SO_2 — reducing capabilities and reliability of systems in use; the power, water and reagents employed; operation and maintenance problems; marketing difficulties; and comparison of capital costs. The study also examines promising new experimental systems now being used.

Land Use

HOW TO JUDGE ENVIRONMENTAL PLANNING FOR SUBDIVISIONS: A Citizen's Guide (February 1981)
For the citizen on a community planning board, this guide provides the essential criteria for judging proposed land subdivisions. What are the right questions to pose to prevent poorly planned developments from becoming costly errors to the community? $3.95

THE INSIDER'S GUIDE TO OWNING LAND IN SUBDIVISIONS: How to Buy, Appraise and Get Rid of Your Lot (January 1981)
A manual for lot buyers. What to know about the environmental practices of a developer and the investment and residential value of a lot—before buying. Also the steps to take if you have already bought and find your lot not to be what you expected. $2.50

PROMISED LANDS 1: Subdivisions in Deserts and Mountains (October 1976)
A study of ten sites in the Southwest and West describing and evaluating the effects of the U.S. land subdivision industry operations on consumers and the environment. $20.00

PROMISED LANDS 2: Subdivisions in Florida's Wetlands (March 1977)
An analysis and evaluation of the environmental and consumer impact of nine Florida subdivisions. $20.00 (Check availability before ordering)

PROMISED LANDS 3: Subdivisions and the Law (January 1978)
An assessment of the effectiveness of land-sales and land-use laws in protecting the environment and consumers. $20.00

BUSINESS AND PRESERVATION: A Survey of Business Conservation of Buildings and Neighborhoods (May 1978)
An examination of corporate activities involving preservation and reuse of existing buildings and historic sites, and support of neighborhood redevelopment. $14.00

Occupational Safety and Health

AT WORK IN COPPER, Volumes 1, 2, 3 (April 1979)
A study of conditions at the 16 U.S. copper smelters, identifying worker safety and health risks, defining the best available worker protection techniques, and evaluating company, government and union efforts to protect employees. Set: $70.00

Volume 1 provides the findings of overall industry performance, with recommendations of feasible engineering controls. Explains criteria upon which smelter evaluations are based. $40.00

Volume 2 profiles smelters owned by ARCO, ASARCO, Cities Services, and Inspiration. $20.00

Volume 3 profiles smelters owned by Kennecott, Louisiana Land, Newmont Mining, and Phelps Dodge. $20.00

HOW OSHA ENFORCES THE LAW (January 1982)
The effectiveness of the Occupational Safety and Health Administration in protecting workers in the copper smelting industry. $15.00

Future Publications

ENERGY STORAGE
A study reporting on near-term applications of 5 energy storage technologies, all of which can help utilities to stretch the usefulness of existing power plants. They may postpone or eliminate the need for new generating capacity by storing off-peak electric power for use during hours of peak demand.

PROMISED LANDS 2 UPDATE
An up-to-the-present examination of land subdivision and sales practices in Florida. Some sites will be revisited, some recent sites will be examined for the first time.

INFORM Staff

Executive Director
Joanna Underwood

Director of Research
Robbin Blaine

Public Education Associate/Editor
Richard Allen

Director of Development
Joan Platt

Director of Administration
Viviane Arzoumanian

Project Directors
Mary Ann Baviello
Ruth Gallo
Cynthia Hutton
Kenneth Pollack
Sophie Weber

Research Associate
Robert Gould

Research Assistants
Alexandra Bowie
Caryn Halbrecht
Michael Jacobs

Public Education Assistants
Risa Gerson
Susan Jakoplic

Copy Editor
Mary Ferguson

Administrative Assistants
Denise Ellis
Patricia Holmes
Patricia Jospe
Linda Post
Linda Vinecour

Student Interns
Michael Gergen
Ilene Green
Jonathan Kalb
Marybeth McCleery
Matthew Roberts
Lisa Rosenfield
Carol Steinsapir
Nancy Warren

Consultants
James Cannon, Energy
Howard Girsky, Publicity
Manuel Gomez, Occupational Safety & Health
Perrin Stryker, Editorial
Daniel Wiener, Energy

Board of Directors

Index